T0220420

Engineering Ethics:
Peace, Justice, and the Earth

Second Edition

Synthesis Lectures on Engineering, Technology and Society

Editor
Caroline Baillie, *University of Western Australia*

The mission of this lecture series is to foster an understanding for engineers and scientists on the inclusive nature of their profession. The creation and proliferation of technologies needs to be inclusive as it has effects on all of humankind, regardless of national boundaries, socio-economic status, gender, race and ethnicity, or creed. The lectures will combine expertise in sociology, political economics, philosophy of science, history, engineering, engineering education, participatory research, development studies, sustainability, psychotherapy, policy studies, and epistemology. The lectures will be relevant to all engineers practicing in all parts of the world. Although written for practicing engineers and human resource trainers, it is expected that engineering, science and social science faculty in universities will find these publications an invaluable resource for students in the classroom and for further research. The goal of the series is to provide a platform for the publication of important and sometimes controversial lectures which will encourage discussion, reflection and further understanding.

The series editor will invite authors and encourage experts to recommend authors to write on a wide array of topics, focusing on the cause and effect relationships between engineers and technology, technologies and society and of society on technology and engineers. Topics will include, but are not limited to the following general areas; History of Engineering, Politics and the Engineer, Economics , Social Issues and Ethics, Women in Engineering, Creativity and Innovation, Knowledge Networks, Styles of Organization, Environmental Issues, Appropriate Technology

Engineering Ethics: Peace, Justice, and the Earth, Second Edition
George D. Catalano
2014

Mining and Communities: Understanding the Context of Engineering Practice
Rita Armstrong, Caroline Baillie, and Wendy Cumming-Potvin
2014

Engineering and War: Militarism, Ethics, Institutions, Alternatives
Ethan Blue, Michael Levine, and Dean Nieusma
2013

Engineers Engaging Community: Water and Energy
Carolyn Oldham, Gregory Crebbin, Stephen Dobbs, and Andrea Gaynor
2013

The Garbage Crisis: A Global Challenge for Engineers
Randika Jayasinghe, Usman Mushtaq, Toni Alyce Smythe, and Caroline Baillie
2013

Engineers, Society, and Sustainability
Sarah Bell
2011

A Hybrid Imagination: Science and Technology in Cultural Perspective
Andrew Jamison, Steen Hyldgaard Christensen, and Lars Botin
2011

A Philosophy of Technology: From Technical Artefacts to Sociotechnical Systems
Pieter Vermaas, Peter Kroes, Ibo van de Poel, Maarten Franssen, and Wybo Houkes
2011

Tragedy in the Gulf: A Call for a New Engineering Ethic
George D. Catalano
2010

Humanitarian Engineering
Carl Mitcham and David Munoz
2010

Engineering and Sustainable Community Development
Juan Lucena, Jen Schneider, and Jon A. Leydens
2010

Needs and Feasibility: A Guide for Engineers in Community Projects — The Case of Waste for Life
Caroline Baillie, Eric Feinblatt, Thimothy Thamae, and Emily Berrington
2010

Engineering and Society: Working Towards Social Justice, Part I: Engineering and Society
Caroline Baillie and George Catalano
2009

Engineering and Society: Working Towards Social Justice, Part II: Decisions in the 21st Century
George Catalano and Caroline Baillie
2009

Engineering and Society: Working Towards Social Justice, Part III: Windows on Society
Caroline Baillie and George Catalano
2009

Engineering: Women and Leadership
Corri Zoli, Shobha Bhatia, Valerie Davidson, and Kelly Rusch
2008

Bridging the Gap Between Engineering and the Global World: A Case Study of the Coconut (Coir) Fiber Industry in Kerala, India
Shobha K. Bhatia and Jennifer L. Smith
2008

Engineering and Social Justice
Donna Riley
2008

Engineering, Poverty, and the Earth
George D. Catalano
2007

Engineers within a Local and Global Society
Caroline Baillie
2006

Globalization, Engineering, and Creativity
John Reader
2006

Engineering Ethics: Peace, Justice, and the Earth
George D. Catalano
2006

Engineering Ethics: Peace, Justice, and the Earth, Second Edition
George D. Catalano

ISBN: 978-3-031-00987-7   paperback
ISBN: 978-3-031-02115-2   ebook

DOI 10.1007/978-3-031-02115-2

A Publication in the Springer series
SYNTHESIS LECTURES ON ENGINEERING, TECHNOLOGY AND SOCIETY

Lecture #22
Series Editor: Caroline Baillie, *University of Western Australia*
Series ISSN
Print 1933-3633   Electronic 1933-3641

Engineering Ethics:
Peace, Justice, and the Earth

Second Edition

George D. Catalano
State University of New York at Binghamton

SYNTHESIS LECTURES ON ENGINEERING, TECHNOLOGY AND SOCIETY #22

ABSTRACT

A response of the engineering profession to the challenges of security, poverty and underdevelopment, environmental sustainability, and native cultures is described. Ethical codes, which govern the behavior of engineers, are examined from a historical perspective linking the prevailing codes to models of the natural world. A new ethical code based on a recently introduced model of Nature as an integral community is provided and discussed. Applications of the new code are described using a case study approach. With the ethical code based on an integral community in place, new design algorithms are developed and also explored using case studies. Implications of the proposed changes in ethics and design on engineering education are considered.

KEYWORDS

engineering ethics, models of the natural world, engineering design, engineering education

With gratitude and appreciation for my family and all my two- and four-legged friends and spiritual directors.

Contents

Preface

I have been a professor for nearly 30 years and have taught thousands of students who have pursued careers in engineering. Over the course of the last several decades, I have, as an engineer and an engineering professor, struggled with issues related to the environmental and societal impacts that technology has in the modern world. I have wondered what views of their responsibilities to society and the natural world do students take with them after graduation? Have I given them the tools to make their way in a world in which the natural world is under siege unlike any time before? How will they respond to the poverty and the injustices which dominate so much of our shrinking global society?

Over the course of my career, I have been a faculty member at colleges in the Deep South, the Midwest, and now the Northeast, at large land grant institutions, elite military academies, and small, predominantly liberal arts universities. I have also been a soldier during times of war. Much has changed in engineering education since my formal schooling where we imagined the engineering profession as value free. Today students do not let us get away with such a narrow view of engineering as more and more of them bring to the classroom an awareness of the state of the world's ecosystem as well as poverty and underdevelopment throughout the world. We in engineering can no longer pretend that such issues are for some other profession, not ours.

My sincere hope in writing the present work is to provide a mechanism whereby issues related to the four great challenges that confront us today—security, poverty and underdevelopment, environmental sustainability, and impact upon native cultures—can be discussed in the context of the engineering profession.

When I first wrote this work, I addressed the three challenges that were set out by the Worldwatch Institute, namely security, poverty and underdevelopment, and environmnetal sustainability. Over the course of the last decade, I have realized that the issues surrounding technology and poverty/development are much more complex than I had first considered. It has become clear to me that the manner in which engineering defined its role was very much linked to our preference for a modern Western lifestyle. It has occurred to me that maybe not all cultures view such a transformation as desirable or healthy. As a result I have included a fourth challenge, one that focuses on the proper and desired goals for the engineering profession while considering the impact on the indigenous peoples throughout the world, although herein the discussion is limited to Native Americans.

George D. Catalano
August 2014

Acknowledgments

I am grateful to many who have helped this effort become a reality. My sincerest thanks are extended to my family, my friends, both two-legged and four-legged, and my many students. I am also grateful to the Re-member Organization located on the Pine Ridge Reservation in South Dakota for their efforts on behalf of the Lakota nation and also in awakening me to many of the biases I have had as an engineer. Each has played a part in the development of the ideas that I have put forward. I will remain forever in their debt. Thank you.

Pax et bene.

George D. Catalano
August 2014

CHAPTER 1

Introduction

According to the Worldwatch Institute, in their annual report on the state of the world, we face three interrelated challenges: the challenge of security—including risks from weapons of mass destruction and terrorism—the challenge of poverty and underdevelopment, and the challenge of environmental sustainability [1].

Recently, I have discovered there is a fourth challenge that also merits consideration and exploration, one centered on the relationships among native or indigenous cultures throughout the world and engineering. Often the concepts of development and poverty have very different meanings than we have here in the Western world.

Technology and rapidly accelerating technical advances have played key roles in the creation of these challenges. Thus, engineers and the profession of engineering have much to say as to whether or not the challenges of security, poverty, and sustainability can be successfully met.

To speak of a profession, particularly the profession of engineering, implies the following five characteristics, which are useful in distinguishing professions from nonprofessional occupations [2]. First, entrance into a profession requires a mastery of some set body of knowledge and thus involves an extensive period of intellectual training. Second, the professionals' knowledge and skills are seen as vital to the well being of the larger society. Third, professions typically have a monopoly or near monopoly on the provisions of their particular set of professional services. Fourth, professionals routinely have an unusual degree of autonomy in the workplace. Fifth, professionals claim to be regulated by ethical standards, usually embodied in a code of ethics. It is this last characteristic, the existence of ethical standards set forth in a code of conduct, which is the focus of the present work.

1.1 THE CHALLENGE OF SECURITY

The challenge of security is at the forefront of everyone's attention today, as it has been every day in the United States since the horrific event of September 11, 2001. That terrible tragedy as well as the 2004 terrorist attacks in Beslan in Russia [3], the bombing of trains in Madrid [4] on March 11, 2004, and many other terrorist attacks in Japan, Indonesia, the Middle east, other parts of Europe, and elsewhere have all driven home the fact that we are not adequately prepared to deal with new threats. But better preparation may suggest a different kind of thinking or approach, not the traditional thinking from the past. This may be especially true for the profession of engineering.

In 2004, by one count, 24 significant on-going armed conflicts (1000 or more deaths) raged around the world, with another 38 hot spots that could slide into or revert to war. Armed conflicts have many costs, in addition to the cost in human lives that is reported in the news. The 1999 Report of the UN Secretary-General [5] put the economic costs to the international community of seven major wars in the 1990s, not including Kosovo, at $199 billion. In addition, there are the costs of economic losses to the countries actually at war. Other important areas affected by armed conflict are: eco-terrorism, environmental destruction as collateral damage, and social casualties.

Over the course of the last decade, the United States has been involved in two wars in Iraq and an armed conflict in Afghanistan and Pakistan. The wars begun in 2001 have been tremendously painful for millions of people in Afghanistan, Iraq, and Pakistan, and the United States, and economically costly as well. Each additional month and year of war adds to that toll. The best estimate of all of the war's recorded dead, including armed forces on all sides, contractors, journalists, humanitarian workers, and civilians, is that over 350,000 people have died due to direct war violence, and many more indirectly [6]. Millions of people have been displaced indefinitely and are living in grossly inadequate conditions. The number of war refugees and displaced persons, over 7 million, is nearly equivalent to the entire population of the New York City fleeing their homes. While we know how many U.S. soldiers have died in the wars (nearly 7000), what is not known are the levels of injury and illness in those who have returned from the wars. New disability claims continue to be filed with the Veterans Administration, with 970,000 disability claims registered as of March 31, 2014 [7]. The U.S. costs for the Iraq war, including an estimate for veterans' medical and disability costs into the future, is more than $2 trillion dollars. The cost for both Iraq and Afghanistan/Pakistan is approaching $4 trillion, not including future interest costs on borrowing for the wars [8].

Consider the impact of wars on children alone. A report prepared for the United Nations General Assembly reveals the full extent of children's involvement in the 30 or so armed conflicts raging around the world [5]. "Millions of children are caught up in conflicts in which they are not merely bystanders, but targets. Some fall victim to a general onslaught against civilians; others die as part of a calculated genocide. Still other children suffer the effects of sexual violence or the multiple deprivations of armed conflict that expose them to hunger or disease. Just as shocking, thousands of young people are cynically exploited as combatants."

According to Vesilind,[1] engineering from its inception has been intimately associated with waging war. The earliest engineers were military engineers who worked at the behest of leaders who either were leading conquering armies or defending their conquered lands from invasion. Vesilund makes reference to the British linguist Young who in 1914 described the lineage of the word engineer tracing it back to the Latin word *ingenium*, an invention or engine [9]. Young adds, "There must have been confusion of Latin *ingenuus* and Latin *ingeniosus*. These should be opposite in meaning. I suppose an engineer ought to be ingenious and ingenuous, artful and

---

[1]P. Aarne Vesilind is an Emeritus Professor of Civil and Environmental Engineering and R.L. Rooke Chair in the Historical and Societal Context of Engineering in the Department of Civil and Environmental Engineering at Bucknell University. At Bucknell, Dr. Vesilind's specialties include environmental engineering and professional ethics, among other areas.

artless, sophisticated and unsophisticated, bond and free." Vesilund concludes with a description of the dichotomy that he claims captures the essence of engineering today:

> "The engineer is sophisticated in creating technology, but unsophisticated in understanding how this technology is to be used. As a result, engineers have historically been employed as hired guns, doing the bidding of both political rulers and wealthy corporations" [10].

1.2   THE CHALLENGE OF POVERTY AND UNDERDEVELOPMENT

According to the Social Summit Programme of Action, "Poverty has various manifestations, including lack of income and productive resources sufficient to ensure sustainable livelihoods; hunger and malnutrition; ill health; limited access or lack of access to education and other basic services; increased morbidity and mortality from illness; homelessness and inadequate housing; unsafe environments; and social discrimination and exclusion. It is also characterized by a lack of participation in decision-making and in civil, social and cultural life. It occurs in all countries: as mass poverty in many developing countries, pockets of poverty amid wealth in developed countries, loss of livelihoods as a result of economic recession, sudden poverty as a result of disaster or conflict, the poverty of low-wage workers, and the utter destitution of people who fall outside family support systems, social institutions and safety nets" [11]. It further emphasizes that "Absolute poverty is a condition characterized by severe deprivation of basic human needs, including food, safe drinking water, sanitation facilities, health, shelter, education and information. It depends not only on income but also on access to social services."

The challenge of poverty and underdevelopment was addressed in October 2004 at the tenth anniversary of the International Conference on Population and Development. The United Nations International Conference on Population and Development (ICPD) was held from September 5–13, 1994 in Cairo, Egypt. During this two-week period, world leaders, high ranking officials, representatives of non-governmental organizations, and United Nations agencies gathered to agree on a Program of Action [12]. The assembly found that significant progress in many fields important for human welfare has been made through national and international efforts. However, the developing countries are still facing serious economic difficulties and an unfavorable international economic environment, and people living in absolute poverty have increased in many countries.

Around the world many of the basic resources on which future generations will depend for their survival and well-being are being depleted and environmental degradation is intensifying, driven by unsustainable patterns of production and consumption, unprecedented growth in population, widespread and persistent poverty, and social and economic inequality. The main conclusions of the final report centered on the fact that while some progress has been made, poverty continues to undermine progress in many areas of the globe. Diseases such as HIV/AIDS are on the rise, creating public health time bombs in many counties. In addition, approximately 20 million children have died of preventable waterborne diseases. Today, hundreds of millions of people continue to live in squalor and destitute conditions associated with the lack of clean drinking water and adequate sanitation.

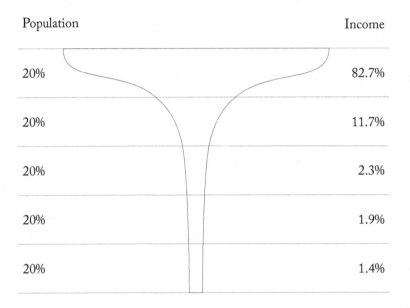

Population		Income
20%		82.7%
20%		11.7%
20%		2.3%
20%		1.9%
20%		1.4%

Figure 1.1: Distribution of world income, 1989 [10].

The report recommends a series of principles [13] that focus on the challenges associated with poverty and underdevelopment in the world. Specifically, Principle 15 states:

"Sustained economic growth, in the context of sustainable development, and social progress require that growth be broadly based, offering equal opportunities to all people."

The following figure (Fig. 1.1) from 1989 UNDP report, shows the unequal distribution of world income and illustrates the underlying reason why 3 billion people have virtually no recourse to the basic necessities of life.

The following table (Table 1.1) from UNDP (United Nations Development Program) shows the millions of people living without the basic necessities of food, education, water, and sanitation.

A fascinating comparison of actual expenditures on luxury items vs. basic needs points to the issue of poverty and underdevelopment as a result of priorities is shown in Table 1.2. Compare the annual expenditure on makeup vs. the funding provided for health care for all women or the money spent on perfumes vs. the funds allocated to universal literacy, for example.

Table 1.1: Priorities of expenditures

ELIMINATING POVERTY: MASSIVE DEPRIVATION REMAINS, 2000 (MILLIONS)

REGION	LIVING ON LESS THAN $1 (PPP US$) A DAY	TOTAL POPULATION UNDER-NOURISHED[a]	PRIMARY AGE CHILDREN NOT IN SCHOOL	PRIMARY AGE GIRLS NOT IN SCHOOL	CHILDREN UNDER AGE FIVE DYING EACH YEAR	PEOPLE WITHOUT ACCESS IMPROVED WATER SOURCES	PEOPLE WITHOUT ACCESS TO ADEQUATE SANITATION
Sub-Saharan Africa	323	185	44	23	5	273	299
Arab States	8	34	7	4	1	42	51
East Asia & the Pacific	261	212	14	7	1	453	1,004
South Asia	432	312	32	21	4	225	944
Latin America & the Caribbean	56	53	2	1	0	72	121
Central & Eastern Europe & CIS	21	33	3	1	0	29	..
World	1,100	831	104	59	11	1,197	2,742

[a] 1998–2000.

Source: World Bank 2003a, 2004f; UNESCO 2003; UN 2003.

Table 1.2: Annual expenditure on luxury items compared with funding needed to meet selected basic needs

PRODUCT	ANNUAL EXPENDITURE (IN BILLION DOLLARS)	SOCIAL OR ECONOMIC GOAL	ADDITIONAL ANNUAL INVESTMENT NEEDED TO ACHIEVE GOAL (IN BILLION DOLLARS)
Makeup	18	Reproductive health care for all women	12
Pet food in Europe and United States	17	Elimination of hunger and malnutrition	19
Perfumes	15	Universal literacy	5
Ocean cruises	14	Clean drinking water for all	10
Ice cream in Europe	11	Immunizing every child	1.3

With few exceptions, engineering and the engineering profession have not addressed issues related to poverty and underdevelopment. Notably, Vallero describes engineering as becoming increasingly complex and the responsibilities of engineers to society as changing as society evolves [14]. According to Vallero, the great challenge in engineering is to treat all people fairly and justly although the motivation for doing so arises out of the observation that undesirable land use results in the creation of unhealthy environments which typically the least advantaged parts of society are forced to bear. Thus, the responsibility to the poor is linked to the destruction of the natural environment not out of a sense of responsibility to address their specific conditions.

1.3   THE CHALLENGE OF ENVIRONMENTAL SUSTAINABILITY

The challenge of environmental sustainability results from a wide-ranging list of critical issues. A few examples of the urgency associated with sustainability are described. Concentrations of carbon dioxide, the main global warming gas in the Earth's atmosphere, posted the largest two-year increase ever recorded. Studies note that if the global temperature rises 2–6 degrees as now predicted, up to 35% of the world's species would become extinct by 2050 [15]. The United Nations published a study noting that the number of oceans and bays with "dead zones" of water, so devoid of oxygen that little life survives, has doubled to 146 since 1990 [16]. Two thirds of Caribbean coral reefs are threatened by human activities, including over-fishing and pollution runoff from agriculture. Toxic metals from discarded cell phones threaten both groundwater and he health of recyclers in Pakistan, India, China, and elsewhere [17].

Polar ice caps and mountain glaciers are experiencing rapidly increasing temperature and as a result are melting at an accelerating rate. The Arctic Climate Impact Assessment study released in 2004 estimates that the Arctic Ocean might be ice-free by the end of the 21st century [18]. The ice sheets covering Greenland and Antarctica are seriously weakening and in many instances disintegrating. This disintegration would cause raising ocean levels around the world and concurrently significant flooding of coastal areas.

Over the course of the last decade, a team of U.S. and Canadian researchers said the Bering Sea was warming so much it was experiencing "a change from arctic to sub-arctic conditions" [19]. Gray whales are heading north and walruses are starving, adrift on ice floes in water too deep for feeding. Warmer-water fish such as pollock and salmon are coming in, the researchers reported.

Off the coast of Nova Scotia, ice on Northumberland Strait has been so thin and unstable during the past few winters that thousands of gray seals crawled on unaccustomed islands to give birth. Storms and high tides washed 1500 newborn seal pups out to sea last year.

Focusing on one particular animal, the main natural habitat of the polar bear is under increasing threat as a consequence of the dramatic thinning of the Arctic sea ice. The link between the thinning of the ice and rising temperatures has been discovered by scientists at UCL and the Met Office Hadley Centre for Climate Prediction and Research, whose findings were published in *Nature* [20]. The thinness of the ice covering the Arctic Ocean, approximately 3 m deep, makes it far more vulnerable to longer summers than the glaciers of the Antarctic. A 40% thinning of the ice has occurred since the 1960s. Polar bears rely on the ice to hunt for seals, and its earlier break-up is giving them less time to hunt. Continued decrease in the Arctic's ice cover would also act to increase the effects of global warming in the northern hemisphere by decreasing the amount of sunlight reflected by the ice. It is also believed that the Arctic ice plays a role in the operation of the Gulf Stream, and that this could be disrupted by the continued thinning.

According to a study funded by the Rockefeller Foundation, 75% of the farmland in sub-Saharan Africa is severely degraded and is being depleted of basic soil nutrients at an ominous rate, deepening a food production crisis that already affects 240 million people. The report, Agricultural Production and Soil Nutrient Miningin Africa, found substantial soil decline in every major region of sub-Saharan African, with the highest rates of depletion in Guinea, Congo, Angola, Rwanda, Burundi, and Uganda, where nutrient losses are more than 60 kg per hectare annually [18]. The problem is exacerbated by fertilizer costs that are 2–6 times the world average, limiting the amount that African farmers can afford to buy, and the adoption of non-traditional farming techniques that sap the soil's fertility.

In summary, the world is in the midst of a period of unprecedented and disruptive change. This is particularly evident when examining the health of the world's ecological systems. A host of human forces impinge upon coral reefs, tropical rain forests, and other critical natural systems located around the world. Half the planet's wetlands are gone. Total carbon emissions and atmospheric concentrations of carbon dioxide are both accelerating and 2004 was the fourth warmest year ever recorded. Over the course of the last 120 years, the 10 warmest years have all occurred since 1990.

Other evidence that supports the existence of a significant disruption to the Earth's climate includes the following facts. Global sea level rose about 17 cm (6.7 in) in the last century. The rate in the last decade, however, is nearly double that of the last century [19]. All three major global surface temperature reconstructions show that Earth has warmed since 1880. Most of this warming has occurred since the 1970s, with the 20 warmest years having occurred since 1981, with all 10 of the warmest years occurring in the past 12 years [20, 21]. Even though the 2000s witnessed a solar output decline resulting in an unusually deep solar minimum in 2007–2009, surface temperatures continue to increase. The oceans have absorbed much of this increased heat, with the top 700 m (about 2,300 ft) of ocean showing warming of 0.302°F since 1969 [22, 23]. Both the extent and thickness of Arctic sea ice has declined rapidly over the last several decades [24]. Glaciers are retreating almost everywhere around the world including in the Alps, Himalayas, Andes, Rockies, Alaska, and Africa [25]. The number of record high temperature events in the United States has been increasing, while the number of record low temperature events has been decreasing, since 1950. The U.S. has also witnessed increasing numbers of intense rainfall events [26]. Since the beginning of the Industrial Revolution, the acidity of surface ocean waters has increased by about 30% [27]. This increase is the result of humans emitting more carbon dioxide into the atmosphere and hence more being absorbed into the oceans. The amount of carbon dioxide absorbed by the upper layer of the oceans is increasing by about 2 billion tons per year [28].

1.4   THE CHALLENGE OF NATIVE CULTURES

The challenge of native cultures often results from a lack of knowledge and ultimately a developing arrogance among us in the West. The engineering profession has embraced the rapidly advancing technologies that surround us as an answer to many of the world's maladies, particularly poverty. Yet the very nature of poverty has been cast in a culturally biased and insensitive way. Shenandoah, an elder and clan mother of the Wolf Clan in the Onondaga Nation in Upstate New York, characterized her childhood as one of growing up in richness [28]. By our standards we would judge her circumstance as one of poverty yet she pointed to the existence of a strong family life, the unspoiled nature of the land, and the abundance of both wild and raised foods much more important than any convenience technology could provide. We could dismiss her analysis as biased, innocent, and ultimately flawed or, alternatively, we could re-examine what it means to be poor. That is the basis for the argument here. Too often the notion of developed vs. underdeveloped has in it a bias towards a particular lifestyle which often is not consistent with many native cultures. More than simply not consistent, the lifestyle of the West has had devastating impacts on many indigenous peoples throughout the entire globe but most especially here on Turtle Island, the native peoples name for North America.

Reference has been made to Native American cultures but the point should be made that there is not just one Native American society. In fact, it is estimated that at the time of the European arrival, there may have been over 500 nations [29]. As a result, for the sake of many meaningful observations, our gaze will be sharpened even more and be directed towards one of those 500 nations, the Lakota culture found on the Pine Ridge Reservation near Rapid City, South Dakota.

The Pine Ridge Indian Reservation ("The Rez") is an Oglala Lakota Native American reservation located in the U.S. state of South Dakota. The reservation consists of 3,000 square miles and has a population of approximately 34,000, although that number is somewhat uncertain [30].

The poverty on Pine Ridge can be described in no other terms than third world [31]. It is common to find homes overcrowded, as those with homes take in whoever needs a roof over their heads. Often children sleep on the floor. Many homes are without running water, and without sewers. The unemployment rate is 80–90% with a per capita income of $4,000. Residents experience eight times the United States rate of diabetes, five times the United States rate of cervical cancer, twice the rate of heart disease, and eight times the United States rate of tuberculosis. The alcoholism rate is estimated as high as 80% and perhaps most troubling of all, the overall suicide rate more than twice the national rate while the teen suicide rate is an incredible four times the national rate.

Pine Ridge is the site of several events that marked tragic milestones in the history between the Lakota and the United States government and its citizens. Stronghold Table, located

Figure 1.2: Typical street scene on Pine Ridge Reservation, South Dakota.

in the Badlands National Park, was the location of the last of the Ghost Dances. The U.S. authorities' attempt to repress this movement eventually led to the Wounded Knee Massacre on December 29, 1890 [32]. A mixed group of Lakota and others sought sanctuary at Pine Ridge after fleeing the Standing Rock Agency, where Sitting Bull had been killed during efforts to arrest him. The unarmed families were intercepted by a heavily armed detachment of the Seventh Cavalry (General George Armstrong Custer's infamous unit), which attacked them, killing over 100, mostly women and children. Later, incredulously, many of the soldiers who took part in the massacre were awarded the U.S. Congressional Medal of Honor. A second tragedy occurred at Wounded Knee in 1973 [33]. On the night of February 27, a long caravan of dilapidated old cars carrying dozens of native men and women, members of the American Indian Movement (AIM) and the local Lakota, traveled into Wounded Knee village. Shortly thereafter, police and Federal agents cordoned off the small town which would soon became the stage of a violent standoff. The AIM and local Lakota held out against the U.S. Government for 71 days. By the time the occupiers left, the village had been destroyed, two Native Americans were dead, and a U.S. Marshal was left paralyzed.

Notwithstanding the poverty and the history of the Lakota on the Rez, rather than wishing to become more "American," there is a significant and growing movement spearheaded by leaders in the Native American community to return to important aspects of their own culture and disengage to a greater extent from some aspects of modern Western society. This is not a 21st century effort similar to the grossly and tragically mischaracterized Ghost Dance[2] of the late

---

[2]The Ghost Dance was a new religious movement in the 1800s incorporated into numerous Native American belief systems. According to the prophet Jack Wilson's teachings, proper practice of the dance would reunite the living with the spirits of the

1800s but rather a deliberate attempt by the Lakota to re-discover and re-dedicate themselves to many of their ways that have existed for thousands of years prior to the European Encounter. An example of such an effort is the Summer Mathematics Camp[3] held on the reservation for elementary school children. Here, lessons in mathematics are embedded within the context of Lakota spirituality and ritual.

1.5 OTHER CHALLENGES

There are other challenges surfacing as we head forward into the 21st century that perhaps are a combination of the three challenges described previously. One issue that has recently gained considerable attention deals with the relation between internet service providers and China. The National People's Congress of the People's Republic of China (PRC) has passed an Internet censorship law in mainland China. In accordance with this law, several regulations were made by the PRC government, and a censorship system is implemented variously by provincial branches of state-owned ISPs, business companies, and organizations. The project is known as Golden Shield.

The operations of U.S. Internet companies in China are attracting concern in Congress after years of complaints from free speech and human rights advocates about these firms aiding Beijing's ability to censor content. Trade liberalization has sent China's economy booming, making it an attractive—even essential market for U.S. companies to enter. But China's government has retained tight political controls. China is believed to have the world's most sophisticated network for monitoring and limiting information online. Combined with Beijing's controls on traditional media, this surveillance program has limited domestic debate on issues such as China's human rights record, Tibetan independence, or Taiwan. Hardware and software provided by many major U.S. technology companies have allowed the PRC government to use the internet as a tool for controlling public opinion.

1.6 CONCLUDING REMARKS

The question then becomes how can the engineering profession effectively respond to these three four challenges, that is, to the challenge of peace, of poverty and underdevelopment, and of environmental sustainability and of native cultures? As responses will depend upon the underlying ethical foundation for the engineering profession as described in the ethical codes, the next section of this work will examine engineering codes of conduct and seek to place them in an historical context.

---

dead and bring peace, prosperity, and unity to native peoples throughout the region. Unfortunately, it was ultimately seen as a threat to U.S. control over Native Americans and its existence has been linked to the massacre at Wounded Knee.

[3]Inila Wakin Janis (*Lakota activist*), a leader and member of the Kit Fox Warrior Society and Cheyenne River Lakota. Janis resides on the Pine Ridge Reservation in South Dakota, and is a Co-Founding member of Green Party USA. He lived for two years in a remote section of the Badlands called the Stronghold, protesting the Parks Service's attempted control over sacred Lakota lands.

CHAPTER 2

Engineering Ethics

Engineering applies technical knowledge to solve human problems. More completely, engineering is a technological activity that uses professional imagination, judgment, integrity, and intellectual discipline in the application of science, technology, mathematics, and practical experience to design, produce, and operate useful objects or processes that meet the needs and desires of a client [34]. Today engineering is seen as a profession which refers specifically to fields that require extensive study and mastery of specialized knowledge and a voluntary and abiding commitment to a code of conduct which prescribes ethical behavior.

2.1 HISTORICAL OVERVIEW

The academic discipline of ethics, also called moral philosophy, involves arranging, defending, and recommending concepts of right and wrong behavior [35]. Philosophers today divide ethical theories into three general subject areas (Fig. 2.1): meta-ethics, normative ethics, and applied ethics. Meta-ethics explores the origins of our ethical standards. Normative ethics seeks to provide standards that can govern right and wrong behavior. Applied ethics focus on specific issues such as, for example, the existence of huge nuclear arsenals in the U.S. and the former Soviet Union, the existence of huge debts among countries in the Third World debt, and the treatment of animals on factory farms. In engineering, codes of conduct, developed to regulate the behavior of the practicing engineer, are examples of normative ethics, which shall be the focus of this review.

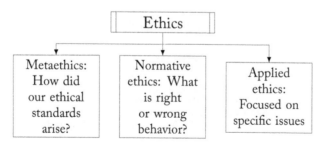

Figure 2.1: Subject areas in ethics.

Normative ethics (Fig. 2.2) involves arriving at moral standards that regulate right and wrong conduct. The Golden Rule [36] is an example of such a moral standard. The key assumption in normative ethics is that there is only one ultimate criterion of moral conduct, whether it is a single rule or a set of principles. Three variations in normative ethics are: (1) virtue theories, (2) duty theories, and (3) consequentialist theories [37].

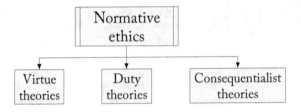

Figure 2.2: Theories of normative ethics.

Virtue ethics places less emphasis on learning rules, and instead stresses the importance of developing good habits of character. Historically, virtue theory is one of the oldest normative traditions in Western philosophy, having its roots in ancient Greek civilization. Plato[1] emphasized four virtues in particular, which were later called *cardinal virtues:* wisdom, courage, temperance, and justice.

Duty theories base morality on specific, foundational principles of obligation. Two terms that are used to describe duty theories are deontological from the Greek word *deon* or duty and nonconsequentialist as these principles are obligatory, irrespective of the consequences that might follow from our actions. Kant's categorical imperative [38] is an example of a duty-based theory of ethics. Kant[2] gives at least four versions of the categorical imperative, but one is especially direct: Treat people as an end, and never as a means to an end. That is, we should always treat people with dignity, and never use them as mere instruments. Ross[3] suggested in a more recent theory that there exists the following duties in living an ethical life [39]:

- fidelity: the duty to keep promises;

- reparation: the duty to compensate others when we harm them;

- gratitude: the duty to thank those who help us;

[1]Plato, 429–347 B.C.E., is considered by most scholars to be one of the most important writers in the Western literary tradition and one of the most penetrating, wide-ranging, and influential authors in the history of philosophy. An Athenian citizen of high status, he displays in his works his absorption in the political events and intellectual movements of his time, but the questions he raises are so profound and the strategies he uses for tackling them so richly suggestive and provocative that educated readers of nearly every period have in some way been influenced by him. In practically every age since the Ancient Greek civilization, there have been philosophers who continue the ideas of Plato and adopt the description as Platonists in some important respects.

[2]Immanuel Kant, 1724–1804, is often considered one of the greatest, and most influential thinkers of modern Europe and the last major philosopher of the Enlightenment. His most original contribution to philosophy is his "Copernican Revolution," that it is the representation that makes the object possible rather than the object that makes the representation possible. This introduced the human mind as an active originator of experience rather than just a passive recipient of perception.

[3]W.D. Ross was a Provost of Oriel College, Oxford, Honorary Fellow of Merton College and a Fellow of the British Academy during the early 20th century. His greatest contribution was found in his criticisms of consequentialist moral theories.

- justice: the duty to recognize merit;

- beneficence: the duty to improve the conditions of others;

- self-improvement: the duty to improve our virtue and intelligence; and

- Non-maleficence: the duty to not injure others.

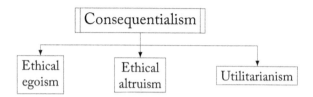

Figure 2.3: Theories of consequentialism.

Consequentialist theories (Fig. 2.3) or principles of ethical behavior require we do a kind of cost-benefits analysis to ultimately decide whether or not an action is ethical or unethical. Consequentialism requires counting or estimating both the good and bad consequences of an action, and determining whether the total good consequences outweigh the total bad consequences. The action is morally proper if the good consequences outnumber the bad consequences. If the bad consequences are greater, then the action is morally improper. Three subdivisions of consequentialism [40] are as follows.

- Ethical Egoism: an action is morally right if the consequences of that action are more favorable than unfavorable only to the agent performing the action.

- Ethical Altruism: an action is morally right if the consequences of that action are more favorable than unfavorable to everyone except the agent.

- Utilitarianism (Fig. 2.4): an action is morally right if the consequences of that action are more favorable than unfavorable to everyone.

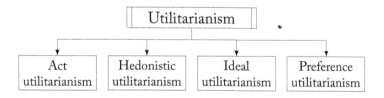

Figure 2.4: Theories of utilitarianism.

It is utilitarianism which has had a significant impact on the development of ethical codes of behavior within the engineering profession as can be observed in a careful reading of the different engineering codes (Fig. 2.5). There have been different variations of utilitarianism devel-

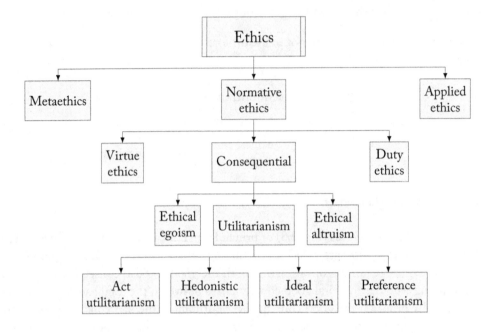

Figure 2.5: Overview of ethics with connections to enginnering codes.

oped. Bentham[4] suggested two kinds: act-utilitarianism (i.e., tally the consequences of each action and thereby determine on a case by case basis whether an action is morally right or wrong) and hedonistic utilitarianism (i.e., only pleasurable consequences matter in the calculation) [41]. Subsequent versions are identified as rule utilitarianism, ideal utilitarianism, and preference utilitarianism. Moore[5] proposed ideal utilitarianism, which involves tallying any consequence that is intuitively recognized as good or bad rather than as causing pleasure or pain [42]. Hare[6] proposed preference utilitarianism, which involves tallying any consequence that fulfills our preferences.

---

[4]Jeremy Bentham, February 15, 1748–June 6, 1832, was an English jurist, philosopher, and legal and social reformer. He is best known as an early advocate of utilitarianism and animal rights, who influenced the development of liberalism.

[5]G.E. Moore, 1873–1959, was a highly influential British philosopher of the early twentieth century. Moore's main contributions to philosophy were in the areas of metaphysics, epistemology, ethics, and philosophical methodology.

[6]Richard Mervyn Hare, March 21, 1919–January 29, 2002, was an English moral philosopher, who held the post of White's Professor of Moral Philosophy at the University of Oxford from 1966 until 1983, and then taught for a number of years at the University of Florida. Peter Singer, known for his work in animal liberation, was also a student of Hare's, and has explicitly adopted many elements of Hare's thought.

2.2   REVIEWING TODAY'S CODES OF ETHICS

At the start of the 21st century, there are as many different codes of conduct in engineering as there are engineering disciplines and specialties. One professional society, the National Society for Professional Engineers (NSPE), has offered one general code which is widely employed today in all the disciplines as well as in engineering education. The NSPE Code of Ethics consists of a preamble followed by a listing of fundamental canons and then rules of practice [43]. The very first canon cautions engineers in the fulfillment of their professional duties, to "hold paramount the safety, health, and welfare of the public." As a result, the first rule of practice states that engineers shall "hold paramount the safety, health, and welfare of the public." Note that the explicit requirements focus on the public though there is no indication as to who is considered to be part of the public. Nor does the code refer to any of the challenges outlined as critical in the previous section. There is no indication that peace and security ought to be considered or issues related to poverty and the underdeveloped world nor environmental sustainability.

The American Society of Mechanical Engineers (ASME) sets forth a similarly constructed code of ethics with fundamental principles followed by fundamental canons [44]. The first principle states that engineers uphold and advance the integrity, honor, and dignity of the engineering profession by using their knowledge and skill for the enhancement of human welfare. The supportive fundamental canon states engineers will hold paramount the safety, health, and welfare of the public in the performance of their professional duties.

The American Society of Civil Engineers (ASCE) does at least mention the environment in its code [45]. According to ASCE, engineers uphold and advance the integrity, honor, and dignity of the engineering profession by using their knowledge and skill for the enhancement of human welfare and the environment (fundamental principle) and will hold paramount the safety, health, and welfare of the public and will strive to comply with the principles of sustainable development in the performance of their professional duties (fundamental canon). There is no explanation of what is meant by the enhancement of the environment. In November 1996, the ASCE Board of Direction adopted the following definition of sustainable development: "Sustainable development is the challenge of meeting human needs for natural resources, industrial products, energy, food, transportation, shelter, and effective waste management while conserving and protecting environmental quality and the natural resource base essential for future development" [46].

The Institute of Electrical and Electronics Engineers (IEEE) Code of Ethics states that its members accept responsibility in making engineering decisions consistent with the safety, health, and welfare of the public, and to disclose promptly factors that might endanger the public or the environment [47]. Here, an interesting notion of responsibility towards the environment is described. It is not in opposition to the IEEE code to endanger the public or the environment only to not disclose promptly factors that might endanger the public or the environment.

The Institute of Industrial Engineers [48] (IIE) endorses the Canon of Ethics provided by the Accreditation Board for Engineering and Technology (ABET) whose first principle is that engineers uphold and advance the integrity, honor, and dignity of the engineering profession by using their knowledge and skill for the enhancement of human welfare and whose first canon is engineers will hold paramount the safety, health, and welfare of the public in the performance of their professional duties [49]. ABET is the accrediting body for all engineering and engineering technology programs in the United States and thus has an important impact on the training of tomorrow's engineers and engineering educators.

Members of the American Institute of Chemical Engineers (AIChE) are challenged to uphold and advance the integrity, honor, and dignity of the engineering profession by being honest and impartial and serving with fidelity their employers, their clients, and the public; striving to increase the competence and prestige of the engineering profession; and using their knowledge and skill for the enhancement of human welfare [50]. To achieve these goals, AIChE members will hold paramount the safety, health, and welfare of the public and protect the environment in performance of their professional duties. There is neither elaboration on the idea of protecting the environment nor an identification of from whom or what will it be protected.

Many other engineering disciplines exist, each with their own codes for ethical conduct. As can be seen from this book, a large percentage of the codes do not explicitly identify the environment as an important stakeholder in discussions of the ethics of engineering choices. Equally as troubling, those codes that do mention the environment refer to the idea of enhancing nature or promoting sustainable development, which is based solely upon meeting human needs. A select few number of codes do mention a responsibility to protect the environment but without identifying from whom or from what. There is no reference to the challenge of peace and security. In addition, there is no reference implicit or explicit to the challenge of poverty and the underdeveloped world. A comparison of many different codes is presented in Table 2.1.

There are many other engineering disciplines at present, each with its own code of conduct or ethics, which describes the responsibilities of the profession. Most focus heavily on the sense of responsibility engineering has towards employers, society in general, and other professional engineers.

2.3   CONCLUDING REMARKS

Engineering as a value-laden profession seeks to codify ethical behavior with various codes of conduct as put forth by different engineering societies. There are differences among the different codes but there are some striking similarities. The similarities exist in what has not been included in the ethical codes. While each does speak to the importance of holding paramount the public safety, issues associated with the intimate connection between engineering and war industries and terrorism are not discussed. In addition, no code speaks to the challenge of world poverty and the plight of the underdeveloped world. With one exception, that of ASCE, the challenge of environmental sustainability is completely ignored.

If we as engineers are to face these important challenges of our time, it may require a significantly different ethical code, one that can only come about if we view our professional

Table 2.1: Comparison of attitudes towards security, poverty, and nature among various engineering societies

CODE OF CONDUCT	RELEVANT CANONS AND PRINCIPLES	SECURITY	ATTITUDES TOWARDS POVERTY	NATURE
NSPE	Hold paramount the safety, health, and welfare of the public.	No explicit reference	No explicit reference	No explicit reference
ASME	Uphold and advance the integrity, honor, and dignity by using their knowledge and skill for the enhancement of human welfare.	No explicit reference	No explicit reference	No explicit reference
ASCE	Hold paramount the safety, health, and welfare of the public and will strive to comply with the principles of sustainable development.	No explicit reference	No explicit reference	Sustainable development linked solely to meeting human needs
IIE	Accept responsibility in decisions consistent with the safety, health, and welfare of the public, and to disclose promptly factors that might endanger the public or the environment.	No explicit reference	No explicit reference	Endangering environment not explored
IIE (ABET)	Will hold paramount the safety, health, and welfare of the public in the performance of their professional duties.	No explicit reference	No explicit reference	No explicit reference
AIChE	Hold paramount the safety, health, and welfare of the public and protect the environment.	No explicit reference	No explicit reference	Protecting the environment not explored

responsibilities in a much broader way. The question then becomes how can we view our sense of ethics in a different way?

The era of modern engineering begin during the Renaissance and flourished as a result of the Industrial Revolution during the 18th and 19th centuries. One of the most important concurrent developments in ethical theory is termed utilitarianism. Utilitarianism (from the Latin *utilis*, useful) is a theory of ethics that prescribes the quantitative maximization of good consequences for a population [51]. It is a single value system and a form of consequentialism and absolutism. This good to be maximized is usually happiness, pleasure, or preference satisfaction. Engineers are by and large utilitarians, seeking to maximum the good that is done [52]. Engineering codes of conduct demonstrate the strong influence of utilitarianism on our sense of responsibility. While utilitarianism has been useful in developing our sense of ethics within engineering, it has not permitted us to broaden that sense of responsibility to include peace and security, the challenge of poverty and the underdeveloped world and of environmental sustainability.

Utilitarianism was developed at a time when the world was imagined to be a machine which ultimately could be analyzed and divided into its many parts. Nature, the Earth, and the Universe were all thought to be governed by immutable laws which we sought to uncover. Yet science has changed dramatically since the era of the mechanical universe. Perhaps we need to examine our sense of ethical responsibility in light of a newer scientific paradigm, one more indicative of the science of the 21st century rather than the 18th and 19th centuries.

CHAPTER 3

Models of the Earth

Various models have been proposed to describe the behavior of the Earth and the Universe. Such models have changed as the scientific paradigm of the time has changed. We will examine several of the models proposed beginning with the Middle Ages in Western Europe and ending with one of the most current models used today at the outset of the 21st century.

3.1 EARTH AS GREAT CHAIN OF BEING

Modern engineering in many respects begin with the Renaissance period in Western Europe. Humankind's understanding of or model for the natural world radically shifted from the notion of the Great Chain of Being [53] prevalent during the medieval period to the Universe as a mechanical clock [54]. The Great Chain of Being, reproduced in Fig. 3.1, is a powerful visual metaphor for a divinely inspired universal hierarchy ranking all forms of higher and lower life; the male alone represents humans. The top of the chain represents perfection in the highest degree. Most believers in the chain call this God. The chain in its entirety represents all degrees of perfection from the highest and fullest to the lowest and least; it is complete. Hence, the universe would not be complete if the chain did not extend all the way to the bottom or if it had gaps in it. The universe is more perfect (in the sense that it is more complete) if all degrees of perfection are represented in it than if only the highest is represented. Eco[1] described this paradigm as "a place for everything and everything in its place" [55].

3.2 EARTH AS MECHANICAL CLOCK

Several voices began to articulate a very different view of the structure of the Universe. One such voice, Giordano Bruno, argued that rather than a transcendent sky God, God was in fact imminent in all things in the Universe [56]. He also argued that rather than one Universe, there were likely an infinite number of Universes or Multiverses as they are referred to today. Galileo is credited with being responsible for one of the most significant revolutions in thought in the development of the Western world and is referred to by many scholars as the originator of experimental science. In fact, the term "revolution" was coined in response to his opinion that our Earth

---

[1]Umberto Eco, born January 5, 1932, is an Italian medievalist, philosopher, and novelist, best known for his novel *The Name of the Rose* and his many essays.

Figure 3.1: Great chain of being a powerful visual metaphor for a divinely inspired universal hierarchy ranking all forms of higher and lower life; the male alone represents humans. From Didacus Valades, Rhetorica Christiana (1579). Reproduced here from Anthony Fletcher's Gender, Sex, & Subordination [53].

was one of several planets that revolved around the sun [57]. The work of Galileo[2] as well as many other natural philosophers led to the Age of Enlightenment, which refers to the 18th century in European philosophy, and is often thought of as part of a larger period, which includes the Age of Reason. Within the "enlightenment" movement, rationality was advocated as a means to establish an authoritative system of ethics, aesthetics, and knowledge. The intellectual leaders of this movement regarded themselves as courageous and elite, and regarded their purpose as leading the world toward progress and out of a long period of doubtful tradition, full of irrationality, superstition, and tyranny which they believe characterized the medieval period. Nature possessed only instrumental value and thus must be not only managed or controlled but transformed into useful resources that fed the insatiable appetites of progress. The visual metaphor used to depict the natural world became the mechanical clock (Fig. 3.2) rather than the Great Chain of Being.

Figure 3.2: Mechanical clock as metaphor for nature (The Tower Clock Cathedral Church of St. James, Toronto).

There was a second revolution, a moral one, which resulted from Galileo's findings. Man was believed to have been made in God's image, was the completion and moral center of the created world. If Galileo was right, man held an insignificant position in the physical universe which did not seems fitting for the moral center of the universe. Physical centrality was understood to signify moral centrality, and Galileo appeared to be denigrating the dignity of man and thereby denying God's scheme of values.

[2]Galileo Galilei, February 15, 1564–January 8, 1642, was an Italian physicist, astronomer, and philosopher who is closely associated with the scientific revolution. His achievements include improvements to the telescope, a variety of astronomical observations, the first and second laws of motion, and effective support for Copernicanism.

The belief that humanity is the moral center of the universe has had lasting endurance. There have been dissenting voices but the predominant view has been that humans are, rightly, of overriding or exclusive moral significance. In our actions regarding the nonhuman world, we have usually only been concerned with human values. Our goals, our technologies, have focused on how best to utilize the natural world to benefit humans. Various religious and secular reasons have been given for our pre-eminent moral standing. Humans unlike the lower animals are said to have souls or to be morally superior because of their rationality. Or humans are said to deserve our privileged position because of our seemingly victorious evolutionary struggle or simply because we have made up the rules.

Consider the attitudes towards the natural world expressed by the various engineering societies as listed in Table 2.1. Clearly, those attitudes originated in the scientific and philosophical theories of the Age of Enlightenment. Engineers should care about nature if at all only if it serves the interests of humankind. Nature has no intrinsic value only instrumental value. Nature needs to be managed, controlled, and manipulated to serve us.

3.3   EARTH AS LIVING SYSTEM

Other metaphors have been used to model the natural world. One which had some notoriety in the late 20th century was the Gaia hypothesis [58]. In 1965, Lovelock[3] published the first scientific paper suggesting the Gaia hypothesis. The Gaia hypothesis, Fig. 3.3, states that the temperature and composition of the Earth's surface are actively controlled by life on the planet [59]. It suggests that if changes in the gas composition, temperature or oxidation state of the Earth are caused by extra-terrestrial, biological, geological, or other disturbances, life responds to these changes by modifying the abiotic environment through growth and metabolism. In simpler terms, biological responses tend to regulate the state of the Earth's environment in their favor.

In support of the Gaia hypothesis, Margulis[4] first put forward her creative theory of endosymbiosis. Endosymbiosis attempts to specify the relationship between organisms which live one within another in a mutually beneficial relationship with one serving as a host cell (the boss cell) and another the symbiont (the dependent organism) which resides within the host cell. As was the case when the Gaia Hypothesis was first outlined by Lovelock, the concept of endosymbiosis was so new and required such a degree of leading edge specialised information, that it was often completely misunderstood—not only by researchers in unrelated fields, but also by her peers.

[3]James Lovelock, born July 26th, 1919, is an independent scientist, environmentalist, author and researcher, He is the author of "The Gaia Theory," and "The Ages of Gaia," which consider the planet Earth as a self-regulated living being.

[4]Lynn Margulis is a Distinguished University Professor at the University of Massachusetts. She has made original contributions to cell biology and microbial evolution to developing science teaching materials and hands-on garbage and trash projects in elementary schools.

Figure 3.3: The Gaia hypothesis: Earth as a living system.

3.4   EARTH AS SELF-ORGANIZING SYSTEM

Modern science at the start of the 21st century does not model the natural world using either the great chain of being or the mechanical clock paradigms or as living being (Gaia hypothesis). Today the natural world is most often described using the model of a self-organizing system and nature rather than being thought of as immutable is seen as constantly in change (Fig. 3.4). Self-organization refers to a process in which the internal organization of a system, normally an open system, increases automatically without being guided or managed by an outside source [60]. Self-organizing systems typically (though not always) display emergent properties. Emergence is the process of complex pattern formation from simpler rules [60]. This can be a dynamic process (occurring over time), such as the evolution of the human brain over thousands of successive

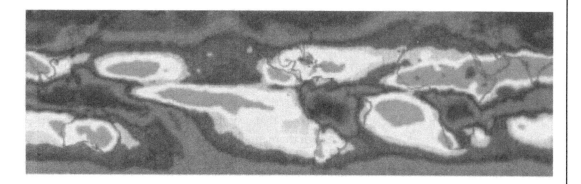

Figure 3.4: An example of a self-organized system. The world's weather patterns (NOAA, December–January 1991) [61].

generations; or emergence can happen over disparate size scales, such as the interactions between a macroscopic number of neurons producing a human brain capable of thought (even though the constituent neurons are not themselves conscious). For a phenomenon to be termed emergent it should generally be unpredictable from a lower level description.

The world abounds with systems and organisms that maintain a high internal energy and organization in seeming defiance of the laws of physics [62]. According to Decker, "As a bar of iron cools, ferromagnetic particles magnetically align themselves with their neighbors until the entire bar is highly organized. Water particles suspended in air form clouds. An ant grows from a single-celled zygote into a complex multicellular organism, and then participates in a structured hive society. What is so fascinating is that the organization seems to emerge spontaneously from disordered conditions, and it does not appear to be driven solely by known physical laws. Somehow, the order arises from the multitude of interactions among the simple parts. The laws that may govern this self-organizing behavior are not well understood, if they exist at all. It is clear, though, that the process is nonlinear, using positive and negative feedback loops among components at the lowest level of the system, and between them and the structures that form at higher levels."

Decker goes on to add, "The study of landscape ecology provides an example of how a self-organizing system or SOS perspective differs from standard approaches. Ecologists are interested in how spatial and temporal patterns such as patches, boundaries, cycles, and succession arise in complex, heterogeneous communities. Early models of pattern formation use a "top-down" approach, meaning the parameters describe the higher hierarchical levels of the system. For instance, individual trees are not described explicitly, but patches of trees are. Or predators are modeled as a homogenous population that uniformly impacts a homogeneous prey population. In this way, the population dynamics are defined at the higher level of the population, rather than being the results of activity at the lower level of the individual."

Decker argues that the problem with this top-down approach is that it violates two basic features of biological phenomena: individuality and locality. By modeling a pack of wolves as a amorphous, homogeneous mob there is no allowance for old wolves, young pups, sick wolves, or aunts and uncles who spend most of their time pack-sitting. Nor is there allowance for expert hunters or, conversely, young and inexperienced hunters.

These differences can lead to larger differences—such as changes in the population gene frequencies, individual body size, or population densities—that might have cascading effects at still higher levels. The principle of locality means that every event or interaction has some location and some range of effect. Decker states, "To say that a system is self-organized is to say it is not entirely directed by top-down rules, although there might be global constraints on the system. Instead, the local actions and interactions of individuals generate ordered structures at higher levels with recognizable dynamics. Since the origins of order in SOS are the subtle differences among components and the interactions among them, system dynamics cannot usually be understood by decomposing the system into its constituent parts. Thus, the study of SOS is synthetic rather than analytic."

If the self-organized system is used to model the natural world rather than the great chain of being or the mechanical clock, our sense of responsibilities to the natural world seem to change significantly. We are forced to look "synthetically" rather than "locally," that is at the very least or moral sphere of concern must broaden. Secondly, nature is no longer in perfect order nor is it a collection of parts (i.e., gears, levers, weight) which can be replaced or modified according to our desires. The mechanical clock in many ways has been replaced by a seemingly chaotic clock which defies predictability, single-valueness and repeatability. If we are to make sense of our place in this natural world, we need a very different sense of ethics. One attempt at providing such an ethical framework has been offered by Johnson in his development of a morally deep world [63].

3.5 CONCLUDING REMARKS

Models that have been proposed to help in our understanding of Nature and the Universe and our place within have changed many times over the course of the last 2,000 years. Nature was imagined once as a great chain of being, with everything in its place and a place for everything. Then with the work of Newton and others, Nature became a machine, losing its enchantment, existing only in material form. Eventually that model was also supplanted by a notion of Nature or the Earth a living organism. Recently, with developments in the science of chaos and complexity, a new model for the Universe has been offered, that is, Nature as a self-organizing system. One thing is clear—there will be more models very soon on our horizon.

CHAPTER 4

Engineering in a Morally Deep World

A new approach to engineering ethics is developed, one based on the notion of a morally deep world. The morally deep world was first developed within the context of environmental ethics. A key element in its development in environmental ethics is the identification of an integral community. The present chapter makes the case for extending the identified integral community to include not only the environment but also other segments of society which have not been included in engineering ethics cases in the past.

4.1 BORROWING FROM ENVIRONMENTAL ETHICS

In A Sand County Almanac [64], Leopold[1] declares: "A thing is right when it tends to preserve the integrity, stability, and beauty of the biotic community. It is wrong when it tends otherwise." According to Leopold, acting ethically is a matter of concern both for us and for others with whom we are in some sort of community. The notion of a community deserves some discussion. We perhaps are most comfortable with community referring to a body of people having common rights, privileges, or interests, or living in the same place under the same laws and regulations; as, a community of Franciscan monks. In biology or ecology, community refers to an interacting group of various species in a common location. For example, a forest of trees and undergrowth plants, inhabited by animals and rooted in soil containing bacteria and fungi, constitutes an integral community. Extending the notion of community in this way is consistent with the pattern evidenced in human society over the centuries. We have progressively enlarged the boundaries of our understanding of community and recognized the membership of slaves, foreigners, etc., those for whom membership was not extended at earlier times in history. Leopold's land ethic then "simply enlarges the boundaries of the community to include soils, waters, plants, and animals, or collectively: the land."

Johnson[2] discusses how non-sentient land can count morally and focuses upon the concept of a living being [65]. For Johnson, a living being is best thought of not as a thing of some sort but

---

[1]Aldo Leopold, 1886–1948, was an American naturalist, conservationist, and philosopher ofprofound importance to the environmental movement. His most significant contribution was the development of the Land Ethic. Leopold wrote, "[A] land ethic changes the role of Homo Sapiens from conqueror of the land-community to plain member and citizen of it. It implies respect for his fellow-members, and also respect for the community as such."

[2]Lawrence Johnson is a philosopher on the faculty at Flinders University, South Australia. He advocates a major change in humankind's attitudes toward the non-human world.

as a living system, an ongoing life-process. A life-process has a character significantly different from those of other processes such as thermodynamics processes for example. Our character, as living beings, is the fundamental determinant of our interests. Johnson adds further that:

> The interests of a being lie in whatever contributes to its coherent effective functioning as an on-going life-process. That which tends to the contrary is against its interests.... moral consideration must be given to the interests of all living beings, in proportion to the interest. Some living systems other than individual organisms are living entities with morally considerable interests. .. . All interests must be taken into account.

A shift to a morally deep world-view in engineering would have a profound impact on the sense of ethical responsibility that the engineering profession would embrace. The next section will provide several examples of application of a morally deep code. Prior to examining those cases, which directly focus upon engineering, I may be useful to consider first a case study involving the environment. The particular case to be examined involves a restoration of a predator, the wolf, into an ecosystem, the southwestern part of the United States.

4.2   CASE STUDY 1: WOLVES IN THE SOUTHWESTERN U.S.

Let us consider the implications of the morally deep argument for the case of restoration of wolves into the Southwest, one of the most contentious issues in wildlife management today [66]. For the purposes of illustration, let us focus on the land near the White Sands Missile Range near Las Cruces, New Mexico. Johnson would challenge us to first identify all the members of the community. For this example a listing would include the following:

- wolves;

- prey animals including domestic sheep and cattle as well as deer, rabbits, coyotes, and others;

- desert lands;

- ranchers and sheep farmers;

- hunters;

- U.S. Fish and Wildlife Service and other state and local government agencies;;

- U.S. Department of Defense;

- residents of White Sands and nearby towns and settlements;

- residents of New Mexico and the entire United States; and

- native American residents.

Often in such cases, two very different perspectives dominate the deliberations. On one side of the debate is atomism, a view that moral assessment applies only to individuals. The individual would be individual wolves, prey, ranchers, etc. On the other side is holism, a view that collectives or whole are subject to moral appraisal. In a morally deep world, the view is shortsighted morally if one adopts either a holistic or atomistic. No one (holistic or atomistic) interest has priority over the other. There is an inevitable tension between atomistic and holistic ethics. Sometime the interests of the biotic community will outweigh the interests of the individual, while at other times it is the interests of the individual, which are paramount. Let us next identify the extent of the community or living being in this case. Recall that a living being is characterized as having an ongoing life process with interests in whatever contributes to its coherent effective functioning. Clearly, wolves, their prey, the desert lands, ranchers, sheep farmers, hunters, and people who live in or near White Sands have considerable interests. Other identified elements could be argued to have less interest in the coherent effective functioning of the community. That is not to suggest that, for example, the residents in New York would have no interest in the restoration but their impact on the coherent effective functioning of the ongoing process would be less.

An interesting example of the tension between atomism and holism can be identified in the following scenario. Suppose wolves are restored to the White Sands Missile range desert and suppose that, as has been the case in Yellowstone National Park, wolves adapt well and quickly grow in numbers [67]. In Yellowstone, some wolves are routinely killed as part of wolf or game management practices [68]. From a holistic perspective this may be morally acceptable but it would be difficult to justify the killing from an atomistic perspective. A morally deep world point of view would argue that both interests need to be considered carefully, including the interests of the entire park community and those of the "surplus" wolf.

A criticism of a morally deep world perspective is that it prevents any action that will affect a community. On the contrary, though a morally deep perspective does assert actions that violate vital interests of the community or erosion of it self-identity should be avoided, it requires active participation in the protection of the essential functions and the maintenance of the viability of life processes. Rather than calling for inaction, a morally deep world perspective suggests contemplation followed by a direct and specific response.

4.3   A NEW ENGINEERING ETHIC

Given a shift to a morally deep world paradigm, a new engineering code of conduct is outlined. The majority of existing codes are structured in similar if not identical ways with fundamental principles supported by fundamental canons. That same structure will be incorporated into the present work. For a morally deep world, the first fundamental canon and rule of practice is specified as:

> *Engineers, in the fulfillment of their professional duties, shall hold paramount the safety, health and welfare of the identified integral community.*

The fundamental difference between an ethical code based on a morally deep world vs. the present codes is the replacement of the "public" by the "identified integral community." The important difference this substitution makes can be seen in the following set of cases.

4.4 CASE STUDY 2: A PLOW FOR MEXICAN PEASANT FARMERS

The case [69] is described as follows: "There is a pressing need for a device to assist third-world peasant farmers in cultivating their small plots of land. This need has never been satisfactorily met by any of the plows currently available. This case involves the design of a plow which can fulfill this need."

The case study went on to identify important questions concerning the significance of the plow.

- For whom should the plow be designed?

- Will humans or animals be used to pull the plow?

- Is the design of the plow sensitive to the gender of the operator?

- Will the operator of the plow walk or ride?

- Should the plow be designed at all?

After careful ethical analysis using present codes, several conclusions are reached. First, it is stated that a designer cannot answer all of the questions posed. In order to do so, the engineer would not only have to do an enormous amount of research, but also have to know the particular social group for which the plow is being designed. Second, quite surprisingly, the stated purpose of the discussion of the case under the present codes did not intend to result in an engineer becoming "obsessed with the cultural and ethical aspects of her (sic) work that she loses sight of more narrowly engineering considerations." Third, the intent of the case study was to raise the issue of "problems of conscience" as they arise in engineering work.

In order to implement a morally deep world code, we will first have to identify members of the integral community. A listing may include:

- peasant farmers;

- landowners;

- animals used in working the land;

- peasant society and culture;

- health of the land;

- indigenous society and culture; and

- local society and culture.

With such identified members of the community, the first two conclusions are very different. Whether or not a great deal of research on the various social groups is no longer an option for the engineer but required. Secondly, leaving aside "obsession," the engineer is required to be aware of the cultural and ethical aspects involve or technical aspects in the proposed design as well as the more narrowly defined engineering

4.5 CASE STUDY 3: A TICKET-TEARING DEVICE FOR A DISABLED PERSON

The case [70] is described as follows: David is a young man who suffers from a variety of physical and mental disabilities. David was employed at a movie theater in his local community near Philadelphia. His primary responsibility was to welcome patrons as they went into the theater hall, taking their admission tickets, tearing them in half and placing the torn tickets into a receiving basket. As David had very limited strength in his hands, the lines of people seeking admittance would soon back up. It was determined by both his employer and social worker that some variety of a device that would help David's pace would be a great aid. A team of senior engineering capstone design students selected this project and dedicated two semesters to the design, fabrication, testing evaluation, and delivery of the final device.

During the two terms, David made several visits to the campus and he and the students became quite close. Delivery day became a highly publicized event with local officials, university officials, family and friends all in attendance along with local and national press. David thoroughly enjoyed the festivities and was immensely pleased by his device. At that time, the project seemed an incredibly successful effort for everyone. Subsequent to the celebration, David continued his work for a while as an attendant at the theater but soon things began to change. He became much more withdrawn than he had ever been and soon quit his job. The seeming depression became worse and worse notwithstanding the heroic efforts of his social worker. David now is completely withdrawn and in fact institutionalized.

For this case, the following questions concerning the significance of the ticket-tearing device can be asked.

- For whom should the ticket-tearer be designed? David? Or a generic client?

- Who will provide technical support for David after the project is finished?

- Who will provide emotional support for David after the project is finished?

- Is the design of the ticket-tearer sensitive to the particular situation of David?

- Is the entire project sensitive to the particular situation of David?

- Should the ticket-tearer be designed at all?

  The identified members of the community may include:

- David;

- David's family;

- technical support workers;

- counseling support;

- theater owner;

- theater patrons; and

- the physically and emotionally disabled.

An objective judging of this case would clearly point to the fact that notwithstanding all the noblest of intentions, David is now worse off than ever before. An engineering team though they followed all appropriate engineering dictums of safety, durability, etc., delivered a device that ultimately may have contributed to the suffering of a young man. What if instead of the engineering codes of conduct and ethics in place today, a morally deep world approach is taken? What would the consequences of such an approach be in this particularly poignant case study?

I would suggest that such an approach would force a much broader consideration of all the factors at play in David's case. There would be consideration given to the impact of not only the device but also the associated attention that the project garnered, an integration of many more professional perspectives, a consideration of not only short term benefits but also those of a much longer time scale. A consideration of David's family and friends and their support for him would be factored into the design.

4.6   CONCLUDING REMARKS

Vesilind [71] in his recent book, *The Right Thing to Do*, rightly points out that the code of ethics is a "fine, first but rough tool for making decisions in engineering." Vesilund also states that the present codes have virtually nothing to say about the environment nor do they spell out what, if any, responsibilities engineers have to non-human animals, plants or places. The present review of existing codes points to the validity of this assertion. A review of presently available case studies for engineering education also suggest a similar struggle with issues associated with social, and cultural impacts. The present work offers one approach to insuring each of these considerations is included in the engineering approach. The approach is based on the notion of a morally deep world first developed within the context of environmental ethics. The key to implementation of this approach is in the identification of key elements in what was referred to as the integral community, that is, both the different elements of a community in an analytical sense but also the community, as a whole, synthetically.

The present codes of ethics used in engineering date back to a period when the universe and all its parts were thought to be nothing more than a great and grand machine. Our understanding of the universe and the natural world here on Earth has dramatically changed since that era. Now, we speak of emergent properties in self-organized systems. I would suggest that it is time for us as engineering educators and engineers to consider adopting a new code of ethics based on our new understanding.

CHAPTER 5

Engineering Design in a Morally Deep World

Engineering design is a process that creates and transforms ideas and concepts into a product definition that satisfies customer requirements. The role of the design engineer is the creation, synthesis, iteration, and presentation of design solutions. The design engineer coordinates with engineering specialists and integrates their inputs to produce the form, fit and function documentation to completely define the product.

An array of different design paradigms are introduced varying from traditional to eco-efficient to eco-effective and eventually including both a basic and enriched morally deep world paradigm. Characterizing each yields the following analysis. The traditional model has been the basis for deisgn in the engineering profession for many years. It focuses on the relationship among the client, public, and profession. An eco-efficient paradigm forces the integration of the possible damaging effects of a design solution on the environment. An eco-effective approach seeks to eliminate the creation of waste. The morally deep world paradigm both strive to significantly broaden the notion of ethical responsibility establishing the notion of an integral community with the enrich model integrating a requirement for self reflection.

5.1 OVERVIEW OF TRADITIONAL ENGINEERING DESIGN

In the traditional modern world view, engineering design and development of new products and services are considered essential for the welfare of most industries in the world [72]. During the past decade, there have been many changes to product and service development but a general outline of the approach is presented in Table 5.1 [73].

Engineers are expected to design any system, subsystem, component or process in accord with the following important design factors: (1) performance; (2) appearance; (3) manufacturing; (4) assembly; (5) maintenance; (6) environment; and (7) safety.

Comparing a traditional engineering design methodology is insightful in light of the challenges of peace and security, poverty and the underdeveloped world and environmental sustainability. While there is a mention of designing for the environment, the methodology explains that "customers and society at large are concerned with preserving the environment" and they also want to minimize the impact of new products and processes upon the environment. There is

Table 5.1: Traditional engineering design algorithm

STEP	DESCRIPTION OF ACTIVITY
1	Identify the customer needs.
2	Establish the product specification.
3	Define alternative concepts for a design that meets specifications.
4	Select the most suitable concept.
5	Design the subsystems and integrate them.
6	Build and test a prototype and then improve it with modifications.
7	Design and build the production facility.
8	Produce and distribute the product.
9	Track the product after release developing an awareness of its strengths and weaknesses.

no mention of issues related to security nor is there any mention of the problem of poverty and the underdeveloped world suggesting that engineering and engineers need not be concerned.

5.2 ECO-EFFECTIVE DESIGN

McDonough[1] and Braungart [74] challenged the notion of minimizing the impact on the environment by asking the provocative question: "When is being "less bad' no good?" They offer a design methodology termed "cradle-to-cradle" based on eco-effectiveness, rather than eco-efficiency (i.e., being "less bad."). The characteristics of a system or product designed according to the different design philosophies are compared in Fig. 5.1.

To illustrate a process or industry which utilizes an eco-effective model of design, McDonough and Braumgart [75] described a community of ants. As part of the ant community's daily activities, they:

- "Safely and effectively handle their own material wastes and those of other species.

- Grow and harvest their own food while nurturing the ecosystem of which they are a part.

- Construct houses, farms, dumps, and food-storage facilities from materials that can be truly recycled.

- Create disinfectants and medicines that are healthy, safe and biodegradable.

- Maintain soil and health for the entire planet."

[1]William McDonough is a world-renowned architect and designer and winner of three U.S. presidential awards: the Presidential Award for Sustainable Development (1996), the National Design Award (2004) and the Presidential Green Chemistry Challenge Award (2003). *Time* magazine recognized him as a "Hero for the Planet" in 1999.

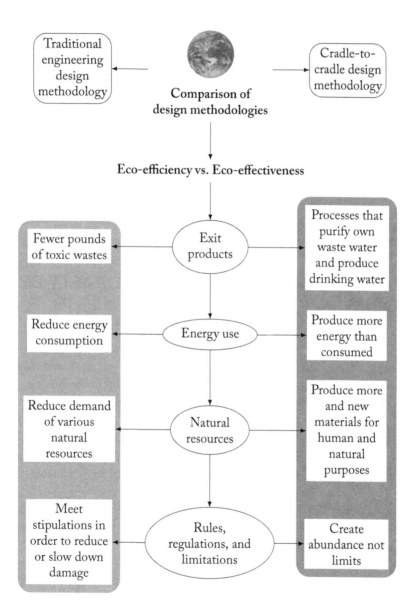

Figure 5.1: Comparison of eco-efficiency vs. eco-effectiveness in design.

The cradle-to-cradle design methodology makes important contributions in promoting environmental awareness and sustainability. This approach is particularly focused on the important issue of respect for diversity and in stark opposition to what the authors' refer to as the "attack of the one size fits all" design response and its attendant de-evolution or simplification on a mass scale. Issues related to the challenge of peace and security and the challenge of poverty and underdevelopment are not addressed directly though there is a reference to producing abundance of natural materials and energy sources. The questions remains—who would have access to this abundance? Would it be distributed throughout the world as wealth is distributed now? How ultimately with that distribution impact questions of peace and security? Such questions seem of pre-eminent importance and need to be more fully addressed.

Engineering within the context of a morally deep philosophy offers the possibility to explicitly intergrate concern both for the health of the environment and for society through its use of the notion of an integral community. A further refinement of the morally deep world design paradigm takes the final step challenging the engineer to reflect upon the values embedded in the proposed solution and if those values are consistent with her/his own hope and a design methodology using the morally deep paradigm will be presented in the following section.

5.3 A DESIGN ALGORITHM FOR A MORALLY DEEP WORLD

A new design algorithm, which can be partitioned in the following four steps, is offered (Fig. 5.2).

- Via Positiva. The problem is identified, fully accepted, and broken down into its various components using the vast array of creative and critical thinking techniques which engineers possess. What is to be solved? For whom is it to be solved?

- Via Negativa. Reflection on the possible implications and consequences for any proposed solution are explored. What are the ethical considerations involved? The societal implications? The global consequences? The effects on the natural environment?

- Via Creativa. The third step refers to the act of creation. The solution is chosen from a host of possibilities, implemented and then evaluated as to its effectiveness in meeting the desired goals and fulfilling the specified criteria.

- Via Transfomativa. The fourth and final step asks the following questions of the engineer: Has the suffering in the world been reduced? Have the social injustices that pervade our global village been even slightly ameliorated? Has the notion of a community of interests been expanded? Is the world a kinder, gentler place borrowing from the Greek poet Aeschylus? [76].

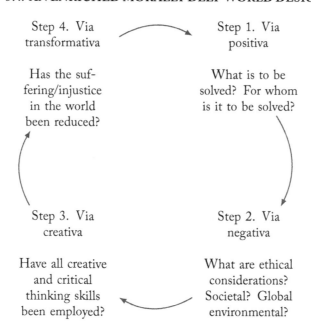

Figure 5.2: Representation of an engineering design algorithm based upon a morally deep world.

5.4   AN ENRICHED MORALLY DEEP WORLD DESIGN ALGORITHM

In this next section, an enhanced design algorithm, consisting of seven steps, using a morally deep world view will be presented. Borrowing from the wisdom traditions of the Lakota, the algorithm visually parallels the Four Directions Prayer as described in the Lakota Medicine Wheel [77]. The Medicine Wheel is a sacred symbol used to represent all knowledge of the universe.

Accordingly, the four directions represent starting with the West, Via Positiva, then proceeding to the North, Via Negativa, then the East direction, Via Creativa and finally on to the South, Via Transformativa. These four steps have been discussed previously and will not be repeated here. That which is different includes the following three steps.: (1) Via Caeleum corresponds to approaching the design problem with an open mind, free from arbitrary or self imagined constraints, cognizant of the large number of possibilities and approaches; (2) Via Terra challenges the designer to consider the design in light of the Earth's health, its present and future impacts; and (3) Via Reflexio challenges the designer to examine his/her own well being in light of the proposed solution, that is, to consider whether or not there exists a unity of the values and purpose associated with the design and the designer's personal values.

- Step 1. Via Caeleum (Looking to the Sky).

White for North representing preservation

PRESERVATION: BODY—skills—maintaining the positive patterns and view of life as an on-going system. Recognizes that Aboriginal people are spirit, heart, mind and body.

Black for West building on your life's lesson

BUILDING: DEVELOPING THE MIND—gaining knowledge, developing the new positive life experiences into continuous patterns and change the view of life which includes integrating the strengths already acquired by the

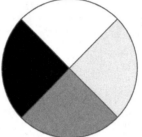

Yellow for East representing awareness

AWARENESS: ATTITUDES AND INSIGHTS into behavioural patterns, ever-increasing understanding of one's self and the world.

Red for South where you pray for your struggles

STRUGGLE: HEART—feelings about self and others and how we interrelate—efforts and attempts to change negative life experiences to positive feeling and believing that my behaviours influence all of my relations.

Figure 5.3: Lakota medicine wheel of life's lessons (http://www.dancingtoeaglespiritsociety.org/medwheel.php).

- Step 2. Via Positiva (Looking to the West).

- Step 3. Via Negativa (Looking to the North).

- Step 4. Via Creativa (Looking to the East).

- Step 5. Via Transformativa (Looking to the South).

- Step 6. Via Terra (Looking towards the Earth).

- Step 7. Via Reflexio (Looking within Oneself).

In the next part of the present chapter, sets of two different cases for the basic as well as the enriched morally deep world design paradigms will be introduced and discussed. For the basic

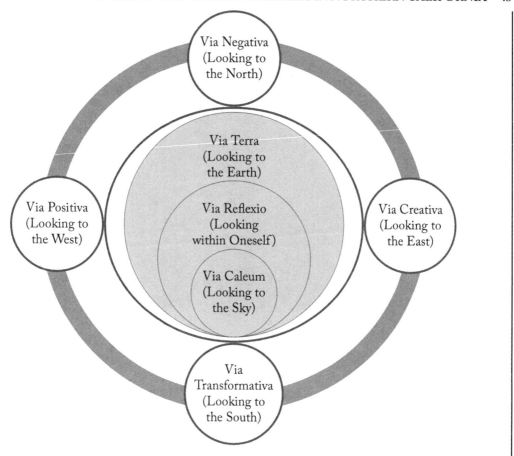

Figure 5.4: Visual representation of enriched morally deep world design paradigm.

paradigm, cases will include harvesting machines in Cailifornia and transporters in the Hudson Bay region in northern Canada. For the enriched paradigm, the cases will include a consideration of hydraulic fracturing in Upstate New York and a recent advancement in bioengineering, germline engineering.

5.5   CASE STUDY 1: GRAPE WORKERS IN NORTHERN CALIFORNIA

Consider the following scenario. Harvesting machines (Fig. 5.5) are replacing migrant workers as grape pickers in northern California [78]. Today, such machines cost approximately $126,000. It takes five workers about eight hours to pick 10 tons of grapes. To harvest the 100 tons would require 90 workers if they wanted to be done before sun-up. The costs are relatively straightfor-

Figure 5.5: An example of a mechanical fruit picker [79].

ward to calculate. It costs $120 a ton to harvest by hand, and $30 a ton by machine. A mid-size wine producer expects to harvest about 4000 tons of grapes total. For 4000 tons, the cost to pick the grapes by hand is approximately $480,000 while the costs using a harvester are $120,000 representing an increase in margin above cost equal to $340,000. Even with the cost of the harvester subtracted from this margin, the net increase in profits is equal to $204,000.

Traditional engineering ethics discussions might stop here. But what if we go a bit farther and identify the workers as being part of the community? What else would we be forced to consider? If it takes 90 workers to harvest 100 tons then it would take the equivalent of 3600 workers to harvest 4000 tons. Certainly there are many more people involved than simply the workers themselves with the total involved growing far beyond the estimate of 3600. What will become of them? Should we as the engineers who designed, built, tested, evaluated and delivered the harvester care? If we use the code of ethics described by countless engineering societies today, the answer would be no. There is no consideration given to the workers whose livelihoods have been eliminated. Codes of engineering ethics based on the notion of a morally deep world would suggest a very different result as it forces us as engineers to consider the entire living system, in this case, including the men, women, and children who toil as grape pickers and whose quality of life is intimately linked to the harvest.

For this case, the following questions concerning the significance of the grape-harvesting device may be asked.

- For whom should the device be designed? The owners? The land? The workers? Someone else?

- What will become of the displaced workers?

- What will become of the displaced workers families?

- Are there long-term effects on the land?

- Are there societal implications for the workers' communities?

- Are there societal implications for the greater community outside the workers and their families?

The identified members of the community may include:

- owners;

- workers;

- workers' families;

- land and water ecosystem;

- farm worker society; and

- local, regional and national societies.

Suppose we use the proposed engineering design algorithm. In Via Positiva, an identification of the problem is made. Grapes have to be picked at a particular time and in a particularly rapid fashion. The harvesting of grapes by hand has in fact been done satisfactorily for thousands of year on farms of all sizes and now to a limited extent by machines. In Via Negativa, the possible implications or consequences of the change to a harvesting machine are considered. The effects upon the environment for the given case seem minimally affected by the mechanization of the process as stated here. The consequences for the workers, their families and their ways of life are however profound and disturbing. A large number of workers will be displaced and their jobs completely eliminated. In Via Creativa, engineers have in fact designed and delivered machines that will meet the criteria set forth by the grape farmers and have thus met their professional responsibilities as has been commonly understood. But a consideration of the net effects on the suffering in the world, Via Transformativa, might yield a different result. Engineers who design such devices without concern for the impact on the numbers of farm workers who are being replaced solely in order to increase the profits of the landowners may have acted unethically.

In the context of the farm workers discussed here, a logical question to ask would be the following: Is it possible to arrive at an end result which was a creative design that met everyone's needs and established justice for workers and their families and furthered the interests of the vineyard? While we cannot offer a device, we can suggest that those whom the new machine would make irrelevant might be included in discussions concerning the consequences of implementing the new design. The landowner may ultimately decide to mechanize this task but at the very least the criteria for making ethical choices would also include a careful consideration of the impact of the engineering design on the lives of the workers. It would seem such an inclusion is as important as any other.

5.6 CASE STUDY 2: TRANSPORTING TOURISTS IN CAPE CHURCHILL

Consider the following case of a dune buggy designer whose clients want a vehicle suitable for taking tourists out along the shore of Hudson Bay to view polar bears. Suppose a company wishes to increase tourism in remote Churchill by promoting the opportunity to view polar bears in an intimate setting. To do this a new version of the ubiquitous dune buggy will be required, one that can withstand the harsh climate, rough terrain, and stand up to an occasional overly curious polar bear. Suppose after much deliberating, the dune buggy's conceptual design is complete. The vehicle will be fully enclosed and heated and be capable of transporting approximately 25 people at a time. Presumably if tourists can view polar bears in this relatively comfortable and safe setting, more tourists will be drawn to Cape Churchill (Fig. 5.6).

For this case, the following questions concerning the significance of the dune buggy device may be asked.

- For whom should the device be designed? The owners? The tourists? The polar bears? Someone else?

- With more tourists, what will be the short-term effects on the local community? The local indigenous community? The polar bears? The tundra?

- With more tourists, what will be the long-term effects on the local community? The local indigenous community? The polar bears? The tundra?

- What are the implications of promoting eco-tourism?

   The identified members of the community may include:

- owners;

- towns people;

- indigenous peoples;

Figure 5.6: Tourists and polar bears interact near Cape Churchill [80].

- polar bears;

- ecosystem in Cape Churchill; and

- local, regional, and national communities.

Whether or not the Arctic-compatible version of a dune buggy is designed and delivered (in this case, it was) would depend upon much more than the needs of a few owners in a morally deep world-view. In the second step of the design process, Via Negativa, attention would be paid to careful considerations of the societal and environmental implications of the device. In the third step of the design process, Via Transformativa, the actual question of whether or not ecotourism is a positive or negative factor for the health of a fragile ecosystem as is the one at Cape Churchill would be raised. In promoting more contact with the polar bears of Cape Churchill are we condemning them and their species to a more rapid extinction? Or are we helping them by bringing a greater awareness to a wider audience?

5.7 HYDRAULIC FRACTURING IN THE MARCELLUS SHALE REGION IN NEW YORK

Hydraulic fracturing has become a contentious political and ethical issue across the nation over the course of the last decade or so. Hydraulic fracturing or "fracking" permits gas and oil deposits typically found in rock formations buried deeply beneath the surface of the Earth. In "fracking," high pressure jets of water coupled with an assortment of chemicals are injected into a drill hole resulting in the fracturing of horizontal drilling the nearby rocks which then permits the escape of the trapped gas and oil. While "fracking" has occurred in the United States since the end of World War II, it has only become economically feasible recently with the development of horizontal drilling and the rise in the price of natural gas.

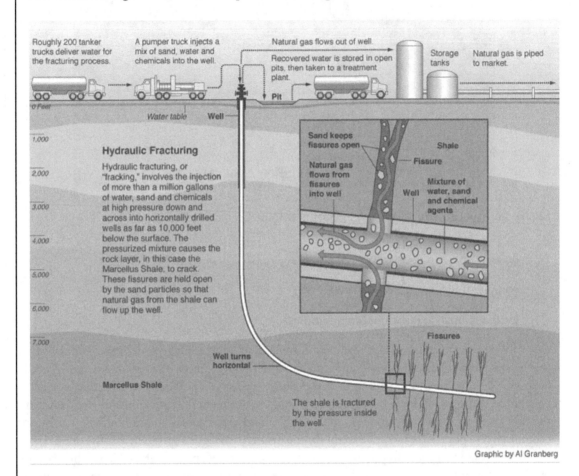

Figure 5.7: What Is hydraulic fracturing? (Graphic by Al Granberg, ProPublica, `https://www.pr opublica.org/special/hydraulic-fracturing-national` [81].)

There are several areas throughout the United States that have developed a "fracking" industry including the Barnett shale formation in Texas, the Haynesville and Fayette shale of Texas, Louisiana, and Arkansas, the Bakken shale of North Dakota, and the Niobrara shale of Colorado, Wyoming, and Nebraska. The present discussion will focus on the Marcellus shale formation located in New York State, Pennsylvania, Ohio, and West Virginia.

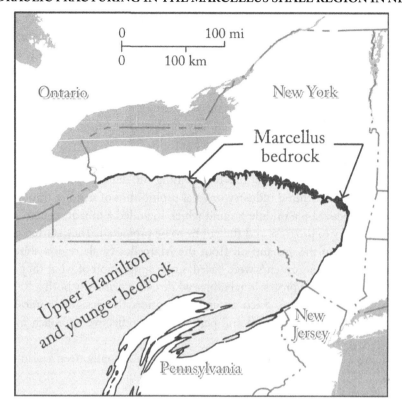

Figure 5.8: Marcellus bedrock formation (Graphics by Dhaluza, `http://en.wikipedia.org/w iki/Marcellus_Formation#mediaviewer/File:Marcellus_Bedrock.svg` Marcellus Bedrock [82]).

During the fracking process water and other chemicals previously mentioned are pumped at high pressures into the well to fracture the rock. The sheer quantity of water needed is enormous. Chesapeake Energy Corporation estimate a single well being hydraulically fractured may consume up to 5 million gallons of fresh water [83]. In addition to the original chemicals of the fracturing fluid, the back flow water is laden with salt and heavy metals such as barium, strontium, and radium [84]. Once the fracturing process is complete, the issue then becomes the disposal of 5 million gallons of back flow. The wastewater must be treated in some way to make the water safe for return to the environment. Other issues include the increase in heavy road traffic and the strain that the increase industrial and business activity will have upon the often times rural communities. A last issue centers on landowners and mineral rights. In many cases, banks will not lend money towards the purchase of land where the mineral rights have been previously leased. This, in effect, locks out the landowner from selling their property to a buyer who would need a traditional loan.

Geologists estimate there may be as much as 489 trillion cubic feet of natural gas—400 times what New York State uses in a year—throughout the Shale. Yet the technique is not without controversy [85]. The group http://business-ethics.com/2012/10/20/10312/www.earthjustice.org has listed many examples of tainted drinking water, polluted air and industrial disasters caused or exacerbated by "fracking" at or near extraction sites since operations began six years ago [86]. A new troubling possible consequence of "fracking" has been found in the Ohio region of the Marcellus Shale, namely earthquakes, and it is also linked as a possible cause for earthquakes in Texas and Oklahoma [87].

Potter and Rashid examined the ethics of hydraulic fracturing [88]. Their approach was to first state the two most extreme solutions for deciding the debate over "fracking," that is, either total deregulation of the drilling industry or total prohibition of drilling throughout the region. They then set out to develop a middle ground which included a mix of regulations and strategies that allow the industry to progress and the public to be protected. They started from the position that the development of the gas and oil from the Marcellus Shale region should be developed to its full potential. Their arguments were based on an application of what they termed common morality, personal ethics and professional ethics while using primarily both a Respect for Persons and Utilitarian ethics paradigm. Such an approach offers the designer some insights into the ethical dilemmas at hand. A different and perhaps more reflective approach might be found in the enriched morally deep world paradigm.

Developing a list of questions keeping the enriched morally deep world design paradigm in mind yields.

- What is the problem we are being asked to solve? Is it energy shortage? Economically driven? Are there other possibilities to consider in either or both cases?

- Is the use of high pressure water jets the only way to crack rock formations? Are there other ways to go about it? What chemicals are added to the high pressure water jets?

- What are the short-, medium-, and long-term effects on the stability of the rock formations?

- What are the short-, medium-, and long-term effects on quality of the ground water?

- How likely are earthquakes? Sink holes?

- How much water will be used? Recycled? Wasted? Is it necessary to use water at all?

- How many jobs will be created in the short term? Medium term? Long term? And what kind of local impact will the jobs have on the local community?

- What will be the impact of drilling on the infrastructure? Who will pay for any repairs?

- What will be the impact on the local ecosystems? Indigenous species?

- What will be the impact of the culture of the local community?

- How do you, the designer, feel about the entire process? Is it consistent with your view on our responsibilities to the environment and the natural world? Water is perhaps the most precious of all natural resources so how does using water in this context mesh with your value system?

- Eventually the economic benefits for the local community will diminish yet the nature of the local community will be likely much different. Does that matter to you?

The elements of the integral community may include:

- landowners and associations;

- local community, its citizens and its governmental agencies;

- local business community;

- gas drillers and support industry employees;

- future generations;

- environmental groups; and

- local and contingent eco-systems.

Let us then consider the ethics of hydraulic fracturing from the perspective of an enriched morally deep world design paradigm. Ultimately, the decision revolves around the relative values that the designer places on professional, economic, environmental, and societal concerns. With this approach, the conflicts, if there are any, are brought to the forefront for careful consideration. For some, the importance of economic self-interest would be paramount while for others, the priceless value of water which can never be artificially created, holds precedence and would preclude using it in this way. Whatever the decision, implications for the environment, for the local community and for the designer personally are not likely to either be ignored or not considered.

5.8   GERMLINE ENGINEERING: A LOOK INTO THE FUTURE NEXUS OF ENGINEERING AND BIOLOGY

The application of engineering analysis and design to living systems holds great promise for the future of humankind but also can present the engineering profession with ethical dilemmas for which it has been inadequately prepared. One approach to dealing with ethical issues located at the interstitial layer between engineering and medical profession based on the morally deep world view will be presented here. In bioengineering or biomedical engineering, a distinction is made between the ethics associated with newly developed devices and techniques in medical practice and the ethics of research and development. In the present discussion focus will be on the former as it is most likely the area in which the most common ethical dilemmas will occur for engineers.

As a minimum, engineering is supportive of the ethical principles that serve as the basis for the practice of medicine [89]. A list of these principles is given below:

- beneficence: Action done for the benefit of others;

- non-maleficence: Actions that do no harm;

- autonomy: Respecting right to self determination;

- dignity: Feeling of comfortableness, in control and valued;

- confidentiality: Obligation to safeguard entrusted information; and

- informed consent: Disclosure of pertinent information with which an individual can make an informed decision.

Medical science has operated both curatively and preventatively. Engineers have historically designed devices and systems to support diagnosis or therapy. With rapid advances in modern technologies, there arises the possibility for engineers to participate in the possibility of designing for human enhancement. Shelly's tale of Dr. Frankenstein and the creation of his monster is no longer that far removed from reality [90].

There are two technologies that warrant discussion: somatic cell therapy and germline engineering. Somatic cell therapy involves the genetic modification of bodily cells other than those found in the sperm or egg cells with its prupose being to replace defective genes with healthy, functional ones [91]. Germline engineering is a more controversial practice in which genes in sperm, egg cells or early life embryos are modified in order to achieve some future outcome [92]. Once these modifications are made, the subsequent modifications to the genome are passed on to future generations ad infinitum. Ethical issues that arise in germline engineering include the following.

Scientific readiness: still remains concerns about the amount and quality of the testing that has taken place and whether or not it is adequate.

Safety concerns: the question is how safe is a gene modification when the future is yet to unfold.

The possibility of "playing God": here deal with the trust we have with future generations and the issue of consent.

Genetic justice: the question is who will be afforded the possibility for enhancement? Will it be based on personal wealth or economic status?

Soon, parents, guardians, agencies of governments or even corporate interests may have the ability to design and deliver what have been termed "super babies." These infants may have advanced physical senses, be more structurally sound or more resistant to infectious diseases or cancers of different types. They may have a longer life expectancy as well, and in addition, greater emotional and/or intellectual intelligence. With all these possible modifications, perhaps the ultimate question that must be raised is what does it mean to be human? Secondly, as an engineer, is it ethically permissible to participate in the development and implementation of technologies that would allow others to "play God?"

The approach taken here is to examine the ethics of germline engineering from the perspective of a morally deep world view.

One of the key elements when undertaking a morally deep world analysis is the development of a list of questions which help clarify the potential ethical dilemmas at hand. In this case, one attempt at developing such a list may include the following questions.

- Is it ethical to change the genetic makeup of an entire line of human descendants? What of the issue of consent? Who will give consent for generations unborn?

- Is there a limit on the number and type of genetic modifications ethically permissible? Are some changes acceptable while others are not? Who would decide?

- Who will have access to the possibility of such modifications? Will access be considered a Universal Human Right?

- Focusing on parents, to what extent ought they be able to change the genes of their own offspring? Moreover, what kind of changes?

The identified members of the integral community may include the following:

- natural parents;

- gestational and/or traditional surrogate parents;

- government agencies;

- corporate interests;

- engineers involved in development and delivery of technical systems, protocols, devices; and

- eco-system and its various species and elements.

Let us then consider the ethics of germline from the perspective of an enriched morally deep world design paradigm. Ultimately, once again the decision revolves around the relative values that the designer places on professional, economic, environmental, and societal concerns.

With this approach, the conflicts, if there are any, are brought to the forefront for careful consideration. For some, the importance of economic self-interest would be paramount while for others, the sacredness and inviolable nature of life holds precedence and would preclude manipulating it in any way. Such manipulations would seem to demand careful consideration of what it even means to be human. Let me be more specific. Suppose it were possible to modify a human embryo's ability to ride bicycles competitively and ultimately win the Tour de France? What does that victory mean with gene manipulation? Is it any different than blood doping? What of all the measures we use as a species to characterize greatness? Who will our new role models be—those who are genetically designed—or are models be those who are not? I am not comfortably morally and ethically with such a brave new world. I would suggest that before a great deal more thought and reflection, neither is our modern world.

5.9 CONCLUDING REMARKS

An engineering design algorithm based on a morally deep world-view has been developed. In fact, two versions have been offered, a basic and what I have chosen to term an enriched model. I have considered a range of cases in a careful examination of the implications of using the new algorithm.

As with all other attempts at integrating an ethical foundation into the engineering profession, it is not the particular ethical theory, which provides a correct answer, as ethical dilemmas do not have right and wrong answers. If such answers existed, we would simply not use the term dilemma—rather the hope is that by providing a range of possibilities, the individual designer will be able to better identify and address the ethical consequences of his/his proposed solutions.

A morally deep world-view integrated into our approach to design does afford us something very important as engineers. It allows us to consider a much broader range of "clients," many more than our simply paying our salaries. It also provides us with a mechanism whereby we can refuse ethically to work on a particular project even though the device itself may meet basic safety requirements. We also may refuse to work on projects that held paramount public safety but excluded considerations of the environment or the impact our device might have upon a community.

CHAPTER 6

Implications for Engineering Education

An engineering profession that defines ethics using a morally deep paradigm would, in turn, require changes to the manner in which students are educated. In the present chapter, an educational paradigm based on the notion of an integral community is described. Engineering education is highly regulated by accrediting agencies, particularly in the United Sates. In many ways, it is the accreditation process that is most effective in promoting change in engineering education. As a result, changes are suggested for the criteria used by the Accreditation Board in Engineering and Technology (ABET) in determining whether or not any program is fully sanctioned.

6.1 A NEW PARADIGM FOR ENGINEERING EDUCATION

In June 2000, an international conference entitled "Connecting Ethics, Ecological Integrity and Health in the Millennium" included one particular presentation, which focused upon what was termed by the authors as a "peace paradigm for education" [92]. The model, the Integral Model of Education for Peace, Democracy and Sustainable Development, adopted from the Earth Charter [93], described three components needed in the shift (Fig. 6.1). The three components describe the three fundamental interrelationships, which form the contexts of our lives: living with ourselves, living with others, and living with nature. Each of these three relationships can be characterized as being marked by violence or by peace. In turn, each of the three relationships can be further subdivided into different dimensions or levels of expression.

6.1.1 LIVING IN PEACE WITH OURSELVES

The interrelationship, living at peace with oneself, is subdivided into levels of expression corresponding to living at peace in the body, heart, and mind. Having peace in our body entails the development of an awareness or consciousness of our physical needs for health and a wise optimization of any/all satisfiers of those needs. This may be seen as an indictment of the "all-nighter" work ethic that is far too common-place among engineering majors. (On a personal note, appealing to students' health and well-being has not been particularly effective for us in teaching time management skills. Perhaps it may be more effective to link the importance of such skills to the promotion of peace.) In addition, being at peace with ourselves certainly points to issues of proper nutrition and substance abuse among others. Peace in our heart requires the meeting of needs that

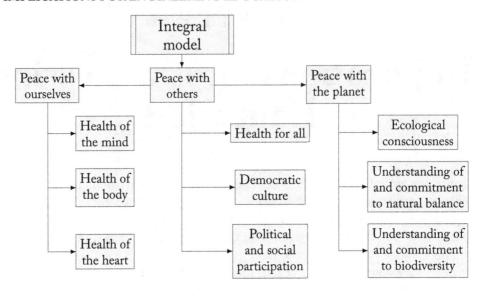

Figure 6.1: Integral model of education for peace, democracy, and sustainable development.

generate a sense of basic security and trust. Here it is necessary to cultivate qualities such as love, compassion, and tolerance. Peace in our mind refers to the possibility of self-realization based on an ethical consciousness of universal responsibility, that is, an appreciation of one's place in natural and human history, and understanding the interdependence of all beings in the universe, as well as the present day global challenges.

6.1.2   LIVING IN PEACE WITH OTHERS

Living in peace with others has the main themes of a truly democratic culture, political and social participation, and health for all. Democratic culture encompasses the key elements of meaningful participation in societal affairs, a sense of responsibility, and a sense of solidarity. Elements involved in political and social participation include a fully engaged and participating citizenship, an identification of the common good, and an understanding of the principles of peaceful conflict resolution. What is not called for is an insistence upon a particular implementation of democracy; rather, there is a recognition and respect that different societies and cultures may reach decisions in a fully democratic way. As an example, we offer the practice of reaching consensus and decisions practiced by our neighbors, the Haudenosaunee, who live on the Onondaga Nation territory in upstate New York. The Onondaga Nation government consists of the traditional Council of Chiefs and Clan Mothers. Additionally, the Onondaga Chiefs sit on the Haudenosaunee Grand Council. Chiefs from each of the Six Nations meet regularly at Onondaga. This governmental organization served as the actual model used by the founders of the American republic [93]. Lastly, the third level or component, health for all, requires an understanding and practice of generosity, an understanding of being itself as a guide for having and doing, and a sense of economic security for all society with an absence of a fear of scarcity.

The translation of living in peace with others to the goals of democracy, political participation, and health may be questioned in the following way: Are these truly universal values or simply the highest values embraced by Western civilization? How can we in the West be sure that we are not arbitrarily imposing these goals on non-Western cultures in order to remake their societies into versions of our own? We would suggest that though these are difficult questions to answer, perhaps the most important contribution we can make as educators is provide a forum within which students may wrestle with these issues.

6.1.3 LIVING IN PEACE WITH THE PLANET

Included in the concept of living at peace with the planet are an ecological consciousness—an understanding and commitment to biodiversity as well as an understanding and commitment to the maintenance of a natural balance. The use of the word "peace" to describe environmental care/stewardship is intentionally provocative. In my view the notion of care and stewardship presupposes a hierarchical relationship between humankind and the planet whereas peace summons forth a metaphor of sitting at a bargaining or "peace" table as equals. In turn, each of these three elements is further subdivided into three components. Ecological consciousness entails identity with the cosmos, an understanding and respect for evolutionary forces and ultimately a respect for life. Here the respect for evolutionary forces does not refer to the narrow view of "survival of the fittest." Rather, it calls on us to recognize that nature is ever changing, ever recasting itself, and not to seek to halt or reverse all such changes. This does not preclude efforts to protect endangered species per se but it does call for a careful consideration prior to intervention. Perhaps, after reflection, society may wish to preserve gray wolves or Bengal tigers through government action while not permitting continued dredging of the Atchafalaya River Basin near New Orleans, Louisiana. Biodiversity consists of an appreciation for the place in the web of life of the various plants and animals, a commitment to the protection of species, particularly endangered species, and a commitment to conservation in concert with the dynamic nature of ecosystems. Natural balance encompasses an appreciation of the integrity of natural systems, an emphasis on sustainable resource use as well as on the importance of ecological security.

6.2 ACCREDITATION CODES AND MODIFICATIONS

In the United States, engineering programs are tightly regulated by the Accreditation Board for Engineering and Technology (ABET) [93]. ABET, Inc., is the recognized U.S. accreditor of college and university programs in applied science, computing, engineering, and technology. ABET was established in 1932 and is now a federation of 28 professional and technical societies representing the fields of applied science, computing, engineering, and technology. According to ABET, their accreditation process ensures the quality of the postsecondary education students receive. Educational institutions or programs volunteer to periodically undergo this review in order to determine if certain criteria are being met. The accreditation is not a ranking system; rather, it is simply assurance that a program or institution meets established quality standards.

Accreditation is a voluntary process on the part of an institution. The steps in the accreditation process are: the engineering program performs a self-study, a team of outside ABET evaluators then visit campus, conduct interviews and compare the program with established criteria, and a final report and finding are returned back to the institution for their review.

ABET sets forth a set of criteria to be met in securing accreditation [89]. The set of requirements cover the quality of students, program educational objectives, programs outcomes and assessment, professional component, the quality of the faculty, classroom, and laboratory facilities, institutional support and financial resources, and program criteria. Considering the existing ABET criteria, Criterion 3 focuses upon program outcomes as shown below. The modified Criterion 3, incorporating the integral model with the changes typed in bold, italics, may be written as follows.

Engineering programs must demonstrate that their graduates have:

- an ability to apply knowledge of mathematics, science, and engineering;

- an ability to design and conduct experiments, as well as to analyze and interpret data;

- an ability to design a system, component, or process to meet desired needs;

- an ability to function on multidisciplinary teams;

- an ability to identify, formulate, and solve engineering problems;

- an understanding of professional and ethical responsibility;

- an ability to communicate effectively;

- the broad education necessary to understand the impact of engineering solutions in a global and societal context;

- a recognition of the need for, and an ability to engage in life-long learning;

- a knowledge of contemporary issues; and

- an ability to use techniques, skills, and modern engineering tools necessary for engineering practice.

The modified Criterion 3 incorporating the integral model would include the following additional program objective (l):

A fully integrative approach to engineering problems incorporating both reason and compassion in the development of solutions.

6.3   CONCLUDING REMARKS

A new model for engineering education incorporating a morally deep world view and the notion of an integral community is presented. The model is based upon making and nurturing peace with ourselves, with others, and with the planet. In addition, as engineering education is highly regulated through the action of accreditation agencies such as ABET, a modification is proposed to the existing ABET Criterion III which supports the integral model by requiring an incorporation of both reason and compassion in the development of proposed technical solutions.

CHAPTER 7

Final Thoughts

I began this discussion with an identification of perhaps the four most pressing challenges we face as a profession in this new millennium: the challenge of security, the challenge of poverty and underdevelopment, the challenge of environmental sustainability, and the challenge of native cultures. The profession of engineering has many important connections to each of the separate challenges. We take great pride in the strong ethical foundation of our profession and have gone to great lengths to try to codify the important ethical dimension of our profession.

Yet, it is here where I think we have fallen short. The ethical codes put forward by countless engineering societies and engineering education agencies are by and large locked into a world-view that was first developed in the Age of Enlightenment. The science of that period in our history was dominated by a view of Nature and the Universe as a grand machine. They, the scientists and philosophers of the time, spoke of Nature as a mechanical clock. As is always the case, science has changed, and now the current paradigm is one of Nature as a self-organizing system with seemingly chaotic systems giving rise to structure through emergent properties. I have offered a different ethical code based upon the science of today as first applied to environmental ethics by Johnson. In his morally deep world-view, Johnson described what he referred to an integral community. It is, I believe, the integration of the notion of an integral community into engineering ethics that holds great promise for the future of our profession.

Consider its implications. When challenged to develop a new device or system, we can now consider a much broader range of "clients" and responsibilities.

To use a visual metaphor, we can now imagine we are dropping a small pebble into a still pond and watching the waves propagate ever outward from the point of impact. We can imagine these waves traveling farther and farther, as far in fact as our sense of responsibility will take us. I would further offer that once we begin to enlarge the area encompassed by that wave, we will be drawn to enlarge it even more.

There is another visual metaphor that may help us in this broadening of ethical responsibility and it arises in many of the engineering sciences. The metaphor is based on the method of characteristics, an analytical solution to the wave equation, one of the fundamental equations in all of physics and engineering. The image can be described in this way.

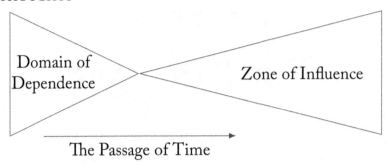

Figure 7.1:

Imagine you are standing in a flowing creek. Your presence in that creek would generate disturbances in the fluid motion (i.e., waves) that would be carried along with the current. The greater the disturbance generated, the wider the wave. In mathematics this would be referred to as the zone of influence. In addition, the speed of propagation may be different than the speed of the current itself. As that propagation speed increases, the zone of influence gets wider and wider. We could turn our gaze back upstream as well and imagine a second zone, in this case referred to as the domain of dependence that describes things that happened in the past that are influencing the present and ultimately the future. Once again, faster rates of propagation would result in a wider domain of dependence. As the rate of change increases, our actions are more widely influenced by what came before us and we influence more widely what happens afterwards. Perhaps all things are connected after all.

As engineers, we have helped develop weapons systems that would only a few years ago seemed more fitting a Jules Verne novel. So too have we helped create incredible wealth for a small portion of the world's community. The impacts of many of our technological advances have been devastating for the world's ecosystems, the species of plants and animals and the indigenous peoples who make up the largest percentage of the world's population. However, as well intentioned as our codes of ethics in the past are, it is my view that they have not met the challenges we have identified for us now in the 21st century nor will they serve us well as we move inexorably into the future.

References

[1] "State of the World 2005," *A Worldwatch Institute Report On Progress Toward a Sustainable Society,* New York: W.W. Norton & Company, 2005, pp. xvii–xxi. 1

[2] Ernest Greenword. "Attributes of a Profession," *Social Work,* pp. 45–55, July 1957. 1

[3] "Psychiatrist Summoned to Moscow to Help Ease Suffering in Breslan, Russia," *Medical News Today,* 10 March 2006, http://www.medicalnewstoday.com/medicalnews.php ?newsid=14033. 1

[4] "BBC News In Depth," http://news.bbc.co.uk/1/hi/indepth/europe/2004/madr idtrainattacks/default.stm. 1

[5] "Militarism and Armed Conflicts," *World Centric,* http://www.worldcentric.org/sta teworld/military.htm. 2

[6] H. McDonald. "Origins of the Word 'Engineer," *ASCE Transactions,* vol. 77, p. 1737. Reproduced in ASCE Committee on History and heritage of American Civil Engineering, 1970, Historical Publication No. 1. 2

[7] Aarne Vesilind. *Peace Engineering,* Woodsville, NH: Lakeshore Press, 2005, pp. 1–2. 2

[8] "Social and Economic Justice," *World Centric,* http://www.worldcentric.org/state world/military.htm. 2

[9] "The Preamble," *Program of Action of the United Nations International Conference on Population & Development,* http://www.iisd.ca/Cairo/program/p01000.html. 2

[10] "Chapter II-Principles," *Program of Action of the United Nations International Conference on Population &Development,* http://www.iisd.ca/Cairo/program/p02016.html. 3, 4

[11] Daniel A. Vallero. "Just Engineering: Peace, Justice, and Sustainability," *Peace Engineering,* Woodsville, NH: Lakeshore Press, 2005, pp. 41–56. 3

[12] Krista E.M. Galley, ed. *Global Climate Change and Wildlife in North America,* Technical Review 04–2, Bethesda, MD: The Wildlife Society, Dec. 2004. 3

[13] AP Wire. *Ocean Dead Zones on the Increase,* CBS News, 10 March, 2006, http://www. cbsnews.com/stories/2004/03/29/tech/main609151.shtml. 4

[14] Lori Brown. "State of the World: A Year in Review," *State of the World 2005,* Norton, New York: Worldwatch Institute, 2005, pp. xxiii–xxvii. 7

[15] Susan Joy Hassol. *Impacts of a Warming Arctic,* New York: Cambridge University Press, Dec. 2004. 7

[16] Brian Handwerk. "Global Warming Could Cause Mass Extinctions by 2050, Study Says," *National Geographic News,* http://news.nationalgeographic.com/news/2006/04/0412_060412_global_warming.html. 7

[17] Jon Copely. "The Great Ice Mystery," *Nature,* vol. 408, pp. 634–636, 07 Dec. 2000. DOI: 10.1038/35047263. 7

[18] "African Farmlands Plagued by Severe Degradation, Study Finds," *The Foundation Directory Online,* The Foundation Center, http://foundationcenter.org/pnd/news/story.jhtml;jsessionid=S3ICALZIOG2OPTQRSI4CGW15AAAACI2F?id=138800010. 7, 8

[19] Church, J. A. and N.J. White (2006), A 20th century acceleration in global sea level rise, Geophysical Research Letters, 33, L01602. DOI: 10.1029/2005GL024826. 7, 8

[20] http://www.ncdc.noaa.gov/oa/climate/research/anomalies/index.html http://www.cru.uea.ac.uk/cru/data/temperature http://data.giss.nasa.gov/gistemp 7, 8

[21] T.C. Peterson et.al., "State of the Climate in 2008," Special Supplement to the Bulletin of the American Meteorological Society, v. 90, no. 8, August 2009, pp. S17-S18. DOI: 10.1175/BAMS-90-8-StateoftheClimate. 8

[22] I. Allison et.al., The Copenhagen Diagnosis: Updating the World on the Latest Climate Science, UNSW Climate Change Research Center, Sydney, Australia, 2009, p. 11. http://www.giss.nasa.gov/research/news/20100121 http://science.nasa.gov/headlines/y2009/01apr_deepsolarminimum.htm 8

[23] S. Levitus, et al., "Global ocean heat content 1955–2008 in light of recently revealed instrumentation problems," Geophys. Res. Lett. 36, L07608 (2009). DOI: 10.1029/2008GL037155. 8

[24] L. Polyak, et.al., "History of Sea Ice in the Arctic," in Past Climate Variability and Change in the Arctic and at High Latitudes, U.S. Geological Survey, Climate Change Science Program Synthesis and Assessment Product 1.2, January 2009, chapter 7. 8

[25] National Snow and Ice Data Center World Glacier Monitoring Service. 8

[26] http://lwf.ncdc.noaa.gov/extremes/cei.html 8

[27] http://www.pmel.noaa.gov/co2/story/What+is+Ocean+Acidification%3F 8

[28] http://www.pmel.noaa.gov/co2/story/Ocean+Acidification 9

[29] A.M. Joseph, 500 Nations: An Illustrated History of North American Indians, N.Y., N.Y. Alfred Kopf. 1994 9

[30] Paul Prucha, Atlas of American Indian Affairs. 9

[31] Pine Ridge Reservation: Re-member. http://www.re-member.org 9

[32] Jack Utter, Wounded Knee and the Ghost Dance Tragedy, Natural Woodlands Pub. Co., April 1991. 10

[33] Peter Matthiesen, In the Spirit of Crazy Horse, Penguin, 1992. 10

[34] "Webster's On-line Dictionary," http://www.webster-dictionary.org/definitio n/ engineering. 13

[35] "Ethics," http://en.wikipedia.org/wiki/Ethics 13

[36] Jeffrey Wattles. *The Golden Rule*, Oxford: Oxford University Press, Dec. 1996. 13

[37] Ethics, *The Internet Encyclopedia of Philosophy*, http://www.utm.edu/research/iep/e /ethics.htm. 14

[38] Kant's Moral Philosophy. *Stanford Encyclopedia of Philosophy*, http://plato.stanford .edu/entries/kant-moral/ 14

[39] W.D. Ross. *The Right and Good*, http://www.ditext.com/ross/right.html DOI: 10.1093/0199252653.001.0001. 14

[40] "Consequentialism," http://www.utm.edu/research/iep/c/conseque.htm 15

[41] Jeremy Bentham. *The Classical Utilitarians: Bentham and Mill*, Indianapolis, IN: Hackett Publishing, Mar. 2003. 16

[42] G.E. Moore. *Principia Ethica*, http://fair-use.org/g-e-moore/principia-ethi ca/ 16

[43] *NSPE Code of Ethics*, National Society for Professional Engineers, http://nspe.org/e thics/ eh-1code.asp. 17

[44] *ASME Code of Ethics of Engineers*, American Society of Mechanical Engineers, http:// asme.org/asmeorg/Governance/5431.doc 17

[45] *ASCE Code of Ethics*, American Society of Civil Engineers, http://www.asce.org/ins ide/codeofethics.cfm 17

[46] *ASCE Code of Ethics*, American Society of Civil Engineers, `http://www.asce.org/ins ide/codeofethics.cfm#note3` 17

[47] *IEEE Code of Ethics*, The Institute of Electrical and Electronics Engineers, `http://www. iee.org/portal/site/mainsite` 17

[48] *IIE Code of Ethics*, The Institute of Industrial Engineers, `http://www.iienet.org/pub lic/` articles/details.cfm?id=79. 17

[49] "2006–2007 Criteria for Accrediting Engineering Programs, Accreditation Board for Engineering and Technology," `http://www.abet.org/LinkedDocuments-UPDATE/Crite riaandPP/E0012006--0720EACCriteria2012-19-05.pdf` 18

[50] *AIChe Code of Ethics*, American Institute of Chemical Engineers, `http://www.aiche.or g/About/Code.aspx` 18

[51] David Hume. *A Treatise of Human Nature (Oxford Philosophical Texts)*, New York: Oxford University Press, 2000. 20

[52] Gail Dawn Baura. *Engineering Ethics, 1st Edition: An Industrial Perspective*, New York: Academic Press, 2006. 20

[53] Christopher Gabbard. *Gender and Politics in Literature, 1688–1750*, `http://www.stanfo rd.edu/class/engl174b/mainpage.html`. 21, 22

[54] Lewis Mumford. *Technics and Civilization*, New York: Harvest/HBJ Books, 1963. 21

[55] Umberto Eco. *The Open Work*, Cambridge, MA: Harvard University Press, 2005. 21

[56] Toby Huff. *An Age of Science and Revolution, 1600–1800*, New York: Oxford University Press, 2005. 21

[57] James E. Lovelock. *Homage to Gaia: The Life of an Independent Scientist*, New York: Oxford University Press, 2001. 23

[58] James E. Lovelock. *Gaia: A New Look at the Earth*, New York: Oxford University Press, 2000. 24

[59] Erich Jantzt. *Self Organizing Universe: Scientific and Human Implications*, New York: Pergamon Press, 1980. 24

[60] F. Eugene Yates, ed. *Self-Organizing Systems: The Emergence of Order*, Springer Verlag, 1988. DOI: 10.1007/978-1-4613-0883-6. 25

[61] Ethan Decker. *Self Organizing Systems: A Tutorial in Complexity*. 25

[62] *Global Radiation Patterns: December–January 1991,* National Oceanic and Atmospheric Administration, `http://www.noaa.gov/` and `http://www.holon.se/folke/images/Globrad1.jpg` 26

[63] Lawrence Johnson. *A Morally Deep World: An Essay on Moral Significance and Environmental Ethics,* Cambridge: Cambridge University Press, 1993, pp. 12–50. 27

[64] Aldo Leopold. *A Sand County Almanac,* New York: Ballantine Books 1986. 29

[65] Lawrence Johnson. *A Morally Deep World: An Essay on Moral Significance and Environmental Ethics,* Cambridge: Cambridge University Press, 1993. 29

[66] Catharine Feher-Elston. *Wolfsong,* New York: Tarcher Press, 2005. 30

[67] Doug Smith and Gary Ferguson. *Decade of the Wolf: Returning the Wild to Yellowstone,* Guilford, CT: The Lyons Press, 2005. 31

[68] Gary Wockner, Gregory McNamee, and Sue-ellen Campbell. *Comeback Wolves,* Boulder, Colorado: Johnson Books, 2005. 31

[69] "Engineering Ethics Case Studies," `http://ethics.tamu.edu/ethics/plow/plow.htm` 32

[70] George D. Catalano. "Compassion Practicum: A Design Experience at the United States Military Academy," *ASEE Journal of Engineering Education,* 2001. DOI: 10.1002/j.2168-9830.2000.tb00553.x. 33

[71] P. Aarne Vesilund. *The Right Thing to Do: An Ethics Guide for Engineering Students,* New Hampshire: Lakeshore Press, 2004. 34

[72] Clive Dym and Patrick Little. *Engineering Design: A Project Based Approach,* New York: Wiley, Nov. 2003. 37

[73] James Daly and David Rocheleau. *Introduction to Engineering Design,* Knoxville, TN: College House Enterprises, LLC, 1998. 37

[74] Michael McDonough and Michael Braungart. *Cradle to Cradle: Remaking the Way We Make Things,* New York: North Point Press, 2002, pp. 45–67. 38

[75] Michael McDonough and Michael Braungart. *Cradle to Cradle: Remaking the Way We Make Things,* New York: North Point Press, 2002, pp. 121–122. 38

[76] Herbert Weir Smith. *Aeschylus,* Cambridge, Massachusetts: Harvard University Press, 1970. 40

[77] Tim Tesconi. "California Grape Harvest 2000," *The Press Democrat,* Oct. 26, 2000. 41

[78] "Harvesting an Olive Orchard," http://www.oliveoilsource.com/harvesting.htm 43

[79] "The Tundra Buggy Adventure," http://www.tundrabuggy.com/polar-bear-tours/cape-churchill/ 44

[80] Leonardo Boff. *Cry of the Earth, Cry of the Poor,* New York: Orbis Press, 1997. 47

[81] Anita Wenden. *Educating for a Culture of Social and Ecological Peace,* New York: State University of New York Press, Nov. 2004. 48

[82] Alexander Ewen and Chief Oren Lyons. *Voice of Indigenous Peoples: Native American People Address the United Nations,* October 1993. 49

[83] *Accreditation Board of Engineering and Technology,* http://www.abet.org 49

[84] *Accreditation Policy and Procedures Manual,* ABET Board of Directors, pp. 2–21. 49

[85] Hydraulic Fracturing, http://www.en.wikipedia.org/wiki/Fracking 50

[86] Earth justice, New York and Fracking, http://www.earthjustice.org/features/newyork-and-fracking 50

[87] Charles Choi, Confirmed: Fracking practices to blame for Ohio earthquakes, NBC News: Science, January 2014 50

[88] A. Potter and M.H. Rashid, An Ethical Approach to Hydraulic Fracturing, 2013 ASEE Southeast Section Conference. 50

[89] Carl Mitcham, Ethics in bioengineering, Journal of Business Ethics, Volume 9, Issue 3, pp. 227–231. DOI: 10.1007/BF00382648. 52, 58

[90] Mary Shelley, Frankenstein, Signet Classics, Reprint edition 2013. 52

[91] Types of Gene Therapy, Gene Therapy Net, http://www.genetherapynet.com/types-of-gene-therapy.html 52

[92] David Heaf, Engineering the Human Germline, http://www.sciencegroup.org.uk/ifgene/germline.htm 52, 55

[93] George D. Catalano, Promoting peace in engineering education: Modifying the ABET criteria, Science & Engineering Ethics, 2006, Volume 12, Issue 2, pp. 399–406 DOI: 10.1007/s11948-006-0039-2. 55, 56, 57

Author's Biography

GEORGE D. CATALANO

George D. Catalano is a Professor of Mechanical Engineering at the State University of New York at Binghamton. He holds joint appointments in the Departments of Mechanical Engineering and Bioengineering. In addition, he serves as a Faculty Master in a university-wide residential communities program at Binghamton. Dr. Catalano earned Doctor of Philosophy and Master of Science degrees in aerospace engineering at the University of Virginia and a Bachelor of Science degree, also in aerospace engineering, at Louisiana State University. Prior to his present position, he served on the faculty at the Air Force Institute of Technology, Wright State University, Louisiana State University, and the United States Military Academy at West Point. He also served as a visiting scholar at the Politechnic in Torino, Italy and at the Technical Institute in Erlangen, Germany.

Dr. Catalano's research interests include turbulent fluid flows, low and high speed aerodynamics and experimental methods in physics, modeling ecosystems, as well as learning strategies and paradigms, engineering ethics, engineering design, and environmental ethics. He is listed in the Philosopher's Index for his published work in animal rights and environmental ethics. Dr. Catalano has over 150 technical and educational publications and has twice been selected as a Fulbright Scholar in recognition of his work in turbulent fluid mechanics.

Printed in the United States
by Baker & Taylor Publisher Services